*Contemporary Physics and
the Limits of Knowledge*

Contemporary Physics and the Limits of Knowledge

Morton Tavel

Rutgers University Press
New Brunswick, New Jersey, and London

Library of Congress Cataloging-in-Publication Data

Tavel, Morton, 1939–
    Contemporary physics and the limits of knowledge / Morton Tavel.
        p. cm.
    Includes bibliographical references and index.
    ISBN 0-8135-3076-8 (cloth: alk. paper) — ISBN 0-8135-3077-6 (pbk. : alk. paper)
    1. Physics. 2. Physics—Philosophy. I. Title.

QC21.3 .T38 2002
530'.01—dc21

2001048398

British Cataloging-in-Publication information is available from the British Library.

Manufactured in the United States of America

Contents

Figures

Preface

This book is the outgrowth of a course for nonscience majors that I have taught at Vassar College for many years. For that reason it could be used as a textbook in a similar course, or can simply be read for enjoyment and, I hope, enlightenment.

In the liberal arts tradition, I emphasize features of physics that make it a very human endeavor and serve to build bridges to other intellectual disciplines. To that end, I have taken as a central theme the impact that modern physics has had on refining the limits to what we can and cannot know both intuitively and empirically. The topics in the book—symmetry, special and general relativity, statistical physics, quantum mechanics, and chaos—have each in their own way challenged the belief of classical physicists that knowledge is limitless and progressive. I try, therefore, to both explain the content of these fascinating topics and emphasize just how they have affected the nature of knowledge.

Since the students that take my course are immersed in the arts, humanities, and social sciences, I draw analogies from those fields and make connections to them whenever possible. Although technical and mathematical details have been kept to a minimum, I heed Einstein's admonition that physics should be made as simple as possible, but no simpler. A topic like chaos, for example, is best understood in terms of the mathematical properties of certain models, so I bite the bullet and present them. In a similar way, it is difficult to explain the development of quantum mechanics without discussing at least some of the experiments and observations that demonstrated the inadequacies of classical physics. I can only hope that the somewhat low level of mathematical and technical sophistication required to read the book is not interpreted as symptomatic of a lack of conceptual sophistication.

Acknowledgments

My commitment to physics education traces back to the two finest physics teachers a student could have had, Lawrence Wills of the City College of New York and David Finkelstein, my doctoral advisor at Yeshiva University. My years at Vassar College helped me to define the role of physics in a liberal arts education. I am particularly grateful to my colleague in Vassar's economics department, Steve Rousseas, who took my relativity course and convinced me to write this book. Many of my ideas about the genetic origins of "common sense" and the way we think about the world came from the writings of Gunther Stent, and I can only hope that I have not added too much entropy to their cogency. I also wish to thank my secretary, Madelyn Thomas, for all her help in preparing the manuscript. Finally, there would be no book at all were it not for the many hours my wife, Judy, spent listening to me and helping me refine my ideas and presentation. In a similar vein, I thank my son, Phillip, for pressuring me to just get it done.

*Contemporary Physics and
the Limits of Knowledge*

One

Introduction

The eminent English physicist Sir Arthur Eddington once remarked that the difference between art and physics is that art makes the commonplace unusual whereas physics makes the unusual commonplace. Certainly anyone who has ever taken a course in physics would tend to agree with at least the latter portion of that statement. Much of the physics that we learn—indeed, much of the science that we learn—is taught with the purpose of explaining how things work and how even the most mysterious phenomena of everyday life can be understood in simple terms. One should not belittle this approach, because there are very good historical and pedagogical reasons for teaching science as though it were a cornucopia of simple explanations for complex things. Historically, most of science actually has been motivated by the search for explanations of poorly understood phenomena. Given that history, it is not surprising that present-day science teachers believe the same factors that have always driven scientists to do science should also encourage younger students to learn science. Teachers of all subjects quickly discover that students tend to learn best when their own curiosity drives the learning process. It is for this reason that the explanation of unusual, but relatively commonplace phenomena is a good theme to use when teaching science.

However, there is also a kind of intellectual "dark side" to science. The immediate goal of science is not always to explain some particular phenomenon, nor can the results of scientific investigations always be measured in terms of an increased understanding of the world around us. Nowhere is this more evident than in the period between the middle of the nineteenth century and the present. This period is fascinating because it is characterized by a subtle but quite astounding change in the nature of physics. Not only has the subject matter of physics changed—that is, the kinds of phenomena studied by physicists—but the very nature of physical explanation has also changed. One of the major accomplishments of physics during this period has been to bring into question the limits on what we can know about the world and how we can know it. Applying a term like "progress" to this period might be seriously questioned, since one of the things we have learned is that we can no longer learn as much. Of course, this rather strange state of affairs must not be cast in a negative light. If we are to embark on an enterprise to learn the secrets of nature, it is certainly best if we know the limits on what it is possible to learn. A hunter wouldn't want to waste time outfitting a safari to hunt unicorns if he knew beforehand that unicorns did not exist. Likewise, it is well to know beforehand the kinds of questions it is legitimate to ask about nature and the kinds of answers we can expect to get. Limits on knowledge, if they are known and understood, actually force us to sharpen our wits. I doubt if anyone has phrased this point with more wisdom and wit than the great French philosopher Blaise Pascal, who wrote:

> Knowledge has two extremes which meet; one is the pure natural ignorance of every man at birth, the other is the extreme reached by great minds who run through the whole range of human knowledge, only to find that they know nothing and come back to the same ignorance from which they set out, but it is a wise ignorance which knows itself. (Blaise Pascal, *Pensees*, trans. A. J. Krailsheimer [Harmondsworth: Penguin Classics, 1988], 51)

This is a book about physics. More precisely, it is about a group of topics in physics that are not only extraordinarily exciting in their own right but also have strong connections to many other intellectual disciplines. These topics are symmetry, special and general relativity, statistical physics, quantum physics, and chaos. To professional physicists and mathematicians, they are as modern and relevant as you can get, and each topic is currently an area of active interest and research. Of course, the treatment of these topics in this book cannot be exhaustive, but it will certainly be adequate to bring you up to date on the current state of knowledge in each area and to show you exactly why each is so important.

My approach will be much more conceptual than technical, but I will not completely shield you from mathematics, equations, and computation, since they are a vital part of the language of physics. You will find, as you read along, that I try to make connections between physics and other disciplines, often the arts. To me, this latter connection is a natural one, and by stressing it I can bring out some of the most beautiful parts of physics.

I began this introduction with Eddington's comment on the difference between art and physics. I think that the similarities are also worth a mention. Most important of these is the fact that artists and scientists alike have as a fundamental goal the description of the world they sense around them. Both are describers and interpreters. Both need an audience. They observe an "outer" world, but what they are ultimately describing is an "inner" one. By that I mean that what you see or otherwise sense as being "out there" is really an image formed inside your own mind. If you are an artist, you may be convinced you are painting the bowl of fruit that rests on that table, but what you are really painting is its mental image. The Belgian surrealist artist René Magritte made this same point more eloquently than I, in a painting entitled *La Condition humaine.*

Magritte said of the world:

> We see it as being outside ourselves even though it is only a mental representation of what we experience on the inside. In the same way we sometimes situate in the past a thing which is happening in the present. Time and space thus lose that unrefined meaning which is the only one everyday experience takes into account. (quoted in *Landscape and Memory*, by Simon Schama [New York: Alfred A. Knopf, 1995], 12)

In his painting Magritte superimposes a window and a canvas. The painting on the canvas is identical with the scene through the window, so that it is impossible to separate one from the other. It is only the legs of the easel and the edge of the canvas that enable the viewer to realize that a painted canvas is there at all. For Magritte, the window to the outside world plays the role of the eye in the body. The canvas is the brain's encoding of the eye's vision. However, what really lies beyond the window?

While artists paint or sculpt their mental images, physicists give form to their mental images in much the same way through the symbols and equations of mathematics. Physicists and artists are not so different after all. Both are convinced they are representing "reality" in some form or other. One thing is certain: the mind plays a central role in both their enterprises, and I shall spend some time discussing exactly how the mind affects the way in which we describe nature.

My treatment of all the topics in this book will be shaped by two basic themes that superficially seem unrelated but are actually connected on a very deep level. The first theme is that science does not simply explain reality; it sometimes creates reality. By this, I mean that science has produced such an incredibly effective way of thinking about the world that, in many instances, the scientific view of things has actually "become" the world. I will expand on this theme as we move along in the book and try to point out how we must think about scientific explanations so that we respect them but do not place them on a pedestal and raise them to untouchable heights.

The second theme is that we human beings, because of the nature of our genetic development, are simply not equipped mentally to understand all levels of natural phenomena. As a result, we have created not one but six different realities. One of these realities is the world as we know and sense it, the world of our direct experience. It is the world we actually "live" in. Most of us think that this is the only reality, but I will try to show you that this is simply not so. We can understand this world so well because our minds and bodies have actually developed in it. In a sense, our brains have their present form precisely so that this world can be thought about and conceptualized. There is in fact a very good reason for believing that our entire mental image of the world around us is a "created" image, an image produced by our brains so that we can survive. In this sense we may create even the everyday reality of our lives, as well as the more sophisticated intellectual and scientific realities that I will discuss in great detail.

The other five "worlds" or "realities" contain physical phenomena that are beyond our direct perception. We have not lived in these worlds, nor can we ever. It is a tribute to the persistence of our curiosity and to our intellectual abilities that we are still able to make sense of these worlds, but we have to do some really fancy creating in order to do so. As we continue our studies of the world around us, we bring these realities together as seamlessly as possible and call the result "contemporary physics."

To help guide you through the six realities of contemporary physics, I will often use the image of maps and mapmaking. The planet Earth has been created by extremely complex processes, some of which we still do not fully understand, but human mapmakers have nevertheless been able to provide us with excellent guides for getting around on it. It would be foolish to confuse their maps with the territory the maps describe. Maps, after all, are made of paper and can be folded up and put in your pocket or in the glove compartment of your car. The planet Earth, on the other hand, is a large, solid sphere that moves and evolves quite independently of our activities on it. Although there is only one Earth, there are an incredible variety of maps, each of which serves a different purpose. There are large-scale and small-scale maps. There are topographic maps and contour maps. There are maps of the oceans and maps of land masses. There are different types of projections. It is amazing to consider the degree to which our mental image of the world is a product of the maps we use.

The same is true of physics. The laws and theories of physics are maps of a reality that we will probably never fully understand. These laws and theories are very useful for getting around in that reality, but we must never confuse them with the reality itself. For one thing, the laws and theories are constantly changing, whereas the reality presumably is not. In addition, there are theories for large-scale reality and theories for reality on the microscopic scale, whereas reality itself is presumably unified. For example, physicists have developed a complex and picturesque hierarchy of elementary particles, ranging down to a small group of what are called "quarks" and "leptons," which they consider more fundamental than all the others. My teacher, David Finkelstein, used to say of those fascinating objects called quarks that they were small particles emitted by physicists. This remark never failed to make me laugh, but I also realized the deep truth that was enclosed in the humor. It could very well turn out that elementary particles have no reality at all other than the fact that they were created by physicists to help them explain reality at some deep inner level.

In the third chapter, I will substantially enlarge upon the second theme, the limitations imposed by the nature of our minds. Let me simply state at this point that I believe our minds are equipped with a set of conceptual structures, which, for lack of a more appropriate phrase, I will lump together and call "common sense." Common sense guides the way we think about the world around us. When coupled with careful observation and logical reasoning, it is more than adequate for developing scientific laws and theories that describe and explain most of the phenomena we are familiar with. These laws and theories then form a map of what most of us call "reality," but which I prefer to call "mesoscopic," or middle-sized, reality. Unfortunately, the term "mesoscopic" has already been appropriated by physicists to refer to a very special size regime, so I will use the more common textbook term "macroscopic" (large-sized) to refer to our own reality.

When common sense fails to guide us accurately, we need new maps that are radically different from those that suffice to describe macroscopic reality. Maps have now been developed to guide us through five new realities, which I call "very fast" (relativistic), "very many" (statistical), "very small" (microscopic), "very large" (cosmic), and "unpredictable" (chaotic). The recognition of the existence of each of these realities (and its corresponding map) is necessitated by the breakdown of one or more of the fundamental common-sense notions that worked for mapping macroscopic reality. My discussion of the limits to common sense will focus on only three of the conceptual structures that, in my opinion, compose it: "cause and effect," "reductionism," and "space and time." Each of these concepts plays a critical role in our understanding of nature.

Before we embark on our project, I will give you some sense of its scope by summarizing the particular challenges that will be raised by the topics of relativity, statistical physics, quantum mechanics, and chaos. These four topics just happen to be the maps of the very fast, the very many, the very small, and the unpredictable. Just as different maps of the world each introduce us to some new way of representing geography, so will each of these maps introduce us to new ideas for thinking about the physical world.

Before we begin relativity, we will first consider the concept of "symmetry," which is the study of changes that cannot be detected. Relativity deals with the question of whether there are certain fundamental features of our own existence that we cannot, in principle, measure or detect. It turns out that there are. Since symmetry is the study of all undetectable changes, it is clear that the concept of symmetry plays a central role in relativity. Symmetry has a counterpart in the mathematical description of the physical world. Whereas symmetry refers to the inability to detect changes in our perception of physical phenomena, the mathematical notion of "invariance" refers to a property of the equations that describe those phenomena. When the phenomena have a symmetry, their equations have an invariance. We always seek the mathematical counterpart to our physical observations. Just as language mirrors thought, so does mathematics mirror the world around us.

When artists view the world, they magnify and even glorify the subjective aspects of their point of view. That is the sense in which they make the commonplace unusual. Physicists, on the other hand, try to eliminate the subjective as much as possible. Physical description dwells on the things that we all agree upon and attempts to eliminate (or carefully control) individual perspectives intruding on our descriptions of nature.

In developing relativity, Albert Einstein carefully examined the role of the observer in producing a description of the natural world. He considered the set of conditions or circumstances in which two (or more) observers are "equivalent"—that is, in which their descriptions of phenomena are essentially identical or at least related in a simple way. We shall see that observers are equivalent when the features of their own particular states of being—that is, their "frames of reference"—play no role in the description of the phenomena they are observing. To obtain a mathematical and physical understanding of this state of equivalence will ultimately require a radical reconceptualizing of the nature of certain phenomena and a careful rethinking of the way we describe them.

Einstein's great contribution was to show that there are certain features of their own states of motion that observers would never be able to measure (which is just what we mean by symmetry), and therefore these features could not and should not play a role in the observers' description of reality. The features do not destroy the equivalence of the observers. If you can't measure it, it mustn't enter into the story you are telling (at least in physics; literature is another matter entirely), and if it is not in your story, it can't be used to distinguish you from any other observer. In short, relativity is a theory of descriptions. More correctly, it is a theory of the relationships between descriptions and the equivalence of descriptions. Since all descriptions ultimately involve language and since the language of physics is mathematics, relativity involves a certain amount of mathematics. I think you will find this enjoyable rather than intimidating.

Whereas relativity leads us to examine the interrelationships among different observers and their descriptions of entire sets of natural phenomena, quantum mechanics deals with the relationship between a single observer and the system under observation. It was developed as an alternative to "classical" or Newtonian mechanics, which was an excellent map of macroscopic reality but proved inadequate to deal with phenomena that were microscopic in scale. To their dismay, physicists discovered that the same laws of physics that so beautifully described the motion of objects ranging in size from planets to apples could not be applied to the domain of molecules, atoms, and nuclei. Somewhere between the apple and the atom, the old theory was failing and a new theory was needed. Quantum mechanics became that theory and is now our map of microscopic reality.

In the process of devising a theory to deal with the world within the atom, it became necessary to scrap the most fundamental assumptions about the kind of description that an observer would be allowed to use. Position and momentum, where an object is located and how fast it is moving (actually, momentum is the product of an object's mass and its velocity), are two essential descriptive quantities of Newtonian mechanics. In the new theory, they could no longer be independently and simultaneously measured with arbitrary precision. This meant that their role in a description of microscopic reality would change substantially. It is not simply that we have one theory for atoms and another for apples; we have one theory for atoms and the things you can say about atoms, and another theory for apples and the things you can say about apples. Newtonian physicists had taken it for granted that their ability to measure position and momentum simultaneously would be carried along to theories about the atomic domain. The reality turned out to be otherwise. Think about constructing a map of the world in which the size and shape of a country and its location on a continent cannot simultaneously be specified!

Toward the end of the nineteenth century, about twenty years before it was discovered that Newtonian mechanics was inadequate to describe atomic behavior, problems with its applicability at even the macroscopic level were beginning to surface. These problems arose from attempts to understand the behavior of large systems of molecules, like solids, liquids, or gases. These difficulties had nothing to do with the size of the molecular components of matter; they arose out of the sheer numbers of molecules that were involved. Thus, quantum mechanics wouldn't have helped, even if it had already been developed and available for use.

At that time, many scientists believed that Newton's mechanical theories should be sufficient to describe the behavior of the sizable collections of individual molecules that composed matter. Although the process might be tedious, it should be a relatively straightforward task to extend Newton's map of reality to include solids, liquids, and gases. In fact, in some simple cases it could be done. But at a more general level, something was going seriously wrong.

Suppose we consider a container filled with some gas as an example. The molecules in the container behave in a simple manner. They either move freely in straight lines (I'll neglect gravity in this example), or else they bump into other molecules or the container walls and move off in different directions, again in straight lines. The free, straight-line motion, as well as the sudden changes produced by collisions, can be correctly described by Newton's mechanics. (These aspects of molecular behavior do not really require quantum mechanics, although that theory is used today for greater accuracy.) What was extremely disturbing was the fact that the behavior of a gas (or liquid or solid) was characterized by a property, called "irreversibility," that Newton's equations could neither predict nor describe. The large system had a "non-Newtonian" property that the Newtonian behavior of its parts couldn't account for. Irreversibility simply means that large collections of molecules inevitably exhibit a kind of "one-way" behavior in time. If you put a drop of cream into your coffee, it eventually spreads out to fill the entire cup. You would never expect to see the cream, once it was mixed into your coffee, gathering itself together and reforming as a single drop. Yet according to Newton's mechanics, as time goes on, individual molecules show no preference for any particular direction in their motions and could, therefore, eventually return to their starting position. If, for example, someone could instantaneously turn every cream molecule around, that should certainly cause the cream to reverse its motion and head back to reform the single drop in the center of the cup. Nevertheless, this sort of thing never happens. In short, the "whole" system does something (in this case, behaves irreversibly) that its molecular "parts" have no reason to do.

This disparity between large-scale and small-scale behavior led to various attempts at reconciliation. Most notable of these was the late-nineteenth-century development, by the Austrian physicist Ludwig Boltzmann, of a theory that started out with Newton's "reversible" mechanics and, rather miraculously, produced the correct irreversible behavior of the gas. Boltzmann's theory created a tremendous controversy and was rejected by some of the most eminent scientists of the day, leading, some say, to his ultimate suicide. While it turned out that Boltzmann did not quite accomplish what he thought he had, he did, in fact, lay the foundations of an entire new theory of molecular behavior, called statistical mechanics. The essence of Boltzmann's idea is that one can get to collective gas behavior from individual molecular behavior by introducing an additional (non-Newtonian) element of probability into the behavior of the molecules. In short, one begins with the precision of Newton's dynamics, then weakens or relaxes that precision by replacing certainty with probability. Probability is a mathematical language peculiarly appropriate for dealing with ignorance. And since ignorance is a theme of this book, it is natural that probability should arise and that we should discuss it.

The final development in modern physics that we will deal with is the theory of what has come to be called "chaos." Since we are in the midst of this development and lack the benefit of historical hindsight, it is less easy to discuss it.

Whereas statistical physics is concerned with describing the complex but predictable behavior that results when an enormous number of molecules each behave in a fairly simple way, "chaos" is the term used to describe a single system that behaves in an enormously complicated and unpredictable way. The notion of chaos arose from the discovery that there are certain apparently simple systems, describable by Newtonian law, whose behavior seems to lack the predictability that characterizes that law. Many things in life are known (or considered) to be unpredictable, but those phenomena usually contain at their core an element of "chance" or "randomness," which we believe accounts for that unpredictability. Chaos, however, is a term reserved for systems that do not have any element of chance in their behavior. Indeed, these systems are describable by reasonably simple equations that can be solved, although not exactly, by use of a computer. What makes their behavior unpredictable is the fact that what they do as time goes on is extraordinarily sensitive to the particular way they are set in motion and that for all practical purposes you can't be sufficiently certain of how they start to be able to predict how they will finish.

Chaos is sometimes referred to as "deterministic unpredictability," a kind of oxymoron. It is "deterministic" in the sense that you can solve the equation that describes the behavior of the system, and therefore any given starting point determines a unique path of subsequent behavior. It is unpredictable in the sense that you can never be sufficiently certain of exactly where the starting point is to make intelligent use of the solution. The existence of such systems seems to be more widespread than we ever thought, which is making the study of chaos quite important.

I would now like to sum up a few of the "things I can't know," according to contemporary physical theory.

1. I can't know my own state of motion (relativity).
2. Although I can predict the general behavior of a complex macroscopic system, I must relate it to the behavior of the constituent microscopic parts by probabilities, because complete knowledge at the microscopic level is impossible to obtain (statistical mechanics).
3. I can't ever know as much about an atomic system as I can about a classical system (quantum mechanics).
4. I can't even predict the behavior of many simple classical systems (chaos).

All this lack of knowledge worries many people. But this is only because they have built up a set of unreasonable expectations about the world based on their many successes in using physics to deal with everyday phenomena. In fact, they have been unbelievably fortunate. In the vernacular, they have "lucked out." In the past they have managed to select those sets of phenomena to describe that are probably the only ones physics gives neat answers to. Maybe it's beginner's luck, like winning a small jackpot with your very first lottery ticket and then expecting success to follow success. It just doesn't work out that way!

In the remainder of this book, we will look more carefully at relativity, statistical mechanics, quantum mechanics, and chaos and discover the beauty in their explanations and predictions. Instead of bemoaning the loss of innocence of Newtonian physics, I would like us to learn to appreciate the world the way it really is.

Two

Science and Creativity

Most people seem to believe that science and art differ in at least one fundamental way. Artists "create," whereas scientists "discover." Had Leonardo da Vinci not painted the *Mona Lisa*, had Michelangelo not sculpted the *Pietà* or had Beethoven not written the Ninth Symphony, none of these "creations" would now exist. If, on the other hand, Newton had not "discovered" the universal law of gravitation, then surely, sooner or later, some other equally observant scientist would have discovered it. This alternative Newton might not have made the discovery as a result of being hit on the head by a falling apple, but no doubt some equally appropriate fruit or vegetable would have come along to play an equivalent role. Gravity, after all, is simply "there," waiting to be found. The laws of nature have already been written. They are in place, merely waiting to be discovered. Works of art, on the other hand, come into being only at the moment of their creation and could clearly not do so in the absence of their creator.

This is a very convincing argument. Nevertheless, it will be a fundamental premise of this book to assert that many of the so-called discoveries of science are, in large measure, just as much creations as are the great works of art, music, and literature. To be sure, some other scientist would have discovered gravity had Newton never been born, but that discovery would have been as different from Newton's as would a painting of a smiling woman by an artist other than Leonardo be different from the *Mona Lisa*.

In order to make my argument convincing, I must clarify what I mean by the terms "creation" and "creativity" when I apply them to science and scientists.

Let me begin by saying what I don't mean. I do not use the words in the common, but limited, sense of personal creativity. Everyone certainly agrees that scientists are creative people. They do what creative people do; they break with tradition and come up with clever, unusual, and innovative ideas and novel ways of looking at old conceptions. No one would deny scientists this personal aspect of creativity. But this is not what I mean. Neither, for that matter, will I make use of a rather obvious corollary of the definition of creativity, that anyone who produces something is its creator and, by definition, is a creative person. I could easily say that the theories put forth by scientists are their creations, simply because theories originate in the minds of scientists, are crafted and polished by their mental processes, and finally are put by their own hands into printed and published form. In this sense, scientists are certainly creative; but, once again, this is not the sense that I am referring to.

I intend to use the word "creation" in a much more radical way than that. I will claim that the very phenomena explained by scientific theories, not just the theories themselves, are creations. The very act of constructing a theory in effect "creates" the phenomenon that the theory then goes on to explain.

Newton did not just create a theory of gravity; he created gravity itself. Of course I do not mean that he created the set of external circumstances that cause apples to fall or our bodies to adhere to the planet Earth. By the same token, when someone says that Leonardo created the *Mona Lisa*, they certainly don't mean that he actually created the young woman whose image appears on the canvas. What I do mean, however, is that Newton created a unique way of thinking about gravity, and he did it so well, so persuasively, and so successfully that his way of thinking about it has become gravity itself. In virtually the same way, "Mona Lisa" is no longer Leonardo's model, it is his canvas. His painting has effectively become, in our minds, Mona Lisa herself. In fact, Leonardo did more than create a painting; he created a "type" and revolutionized portraiture. When we think of the *Mona Lisa*, we envision the woman's enigmatic smile, her flawless complexion. She stands out in front of the minor landscape that Leonardo placed behind her. But all that is Leonardo's creation. None of us knows what the model herself looked like, nor do we even care. She is what Leonardo made her.

And Newton did precisely the same thing. He treated gravity as a force, and so it became a force. We all now think of it as a force. "What is gravity?" I ask a student. "It is the force that pulls us to the earth," the student replies. Yet it wasn't a "force" before Newton treated it as such; moreover, as we shall shortly see, it is no longer a force, thanks to Einstein.

Michelangelo is supposed to have remarked that when he sculpted, he simply removed the unnecessary pieces from a block of marble and revealed the statue that had always been within it. We might be tempted to argue that art, too, is discovery. The Czech poet Jan Skácel wrote a quatrain in which he claimed that poets don't invent poems, they simply find them. Each poem has already been there for a long time. Based on the assertions of artists like Michelangelo and Skácel, an argument could be made for the act of discovery in art and literature. But my task is to convince you of the creating aspect of science rather than the discovering aspect of art. Let me add one note of warning. I do not claim that all science is creation. Neither, for that matter, would I claim that all paintings are creations. In fact, most of science really is discovery; careful observations and laboratory experiments that flesh out and set limits on the broader laws and theories. The truly great works of science, however, are creations, and those are the ones we will deal with in this book.

Since we will spend a great deal of time discussing certain specific creations of Einstein, I wil give a more precise sense of what I mean by creation using the example of gravity. The many effects of gravity have been known for thousands of years (so to that extent it was discovered long before Newton developed his theory), but the explanation of exactly what gravity "is" and how it works has undergone a series of fascinating changes.

To make the point clear, imagine Aristotle, Newton, and Einstein together, observing what is undoubtedly the most common and fundamental of all gravitational phenomena, the motion of a falling object (say, an apple). Each of the three then provides an explanation of why the apple falls. "The reason why that apple falls" says Aristotle, "is quite clear. It falls in order to restore the harmony of nature. When the apple is held above the earth, it is removed from its source, the earth itself. It falls to restore 'wholeness' and thereby to achieve harmony." To Aristotle, gravity is a facet of his philosophy of reality, which supposes that there is a tendency of nature to move toward a condition of harmony, a wholeness in which all things made of the earth's substance are restored to the earth. What then is Aristotle's "discovery" when he discovers gravity? He has in fact already "created" his philosophy of harmony. Gravity is simply one more manifestation of that creation.

Newton observes the same falling apple and probably says something like this: "I have already demonstrated that when massive objects are acted on by forces, and only then, they acquire a particular state of motion which I call 'accelerated motion.' The presence of acceleration can be verified by measurement. Since I can measure the fact that this apple, while it is falling, is in just such a state of accelerated motion, it must be subject to a force. That force must be exerted on the apple by the earth, because I see no evidence of any other reasonable cause. I shall call that force gravity." Newton apparently has "discovered" that gravity is a force between the earth and the apple. But is this actually a discovery? He has already "created" a set of relationships between forces and the motions they produce. He can discover gravity only because he has a creation within which to fit it. In short, Newton can discover a force of gravity only because he has already created the concept of force.

Finally, Einstein has a shot at it. "This apple, like all matter and energy in the universe, must move through spacetime, which is the very fabric of existence. The fact that to us (as Sir Isaac has already noted) it appears to be falling faster and faster toward the earth is merely a manifestation of the fact that spacetime itself is curved in the vicinity of the earth and, among other things, clocks move at different rates in regions of different curvature. What we call gravity is merely our perception of motion through a curved spacetime." Einstein too has "discovered" gravity. He has discovered it to be a perception.

What then is gravity? Is it the seeking of harmony? Is it a force between massive objects? Is it a peculiar perception produced by motion through a curved spacetime? In fact, whichever explanation you choose to believe, none of them are discoveries. They are all creations. The only thing we ever really discover is that an apple, breaking loose from the branch that held it, ultimately strikes the ground. This is basic, raw, visual experience. It is observation. Upon this observation we impose a "reason" or explanation, which is a purely mental construction—that is, a creation. Gravity is the reason; it is not the observation. Leonardo was certainly observing someone as he painted the *Mona Lisa*. The act of observation was a discovery; the act of painting was the "explanation," the "creation."

There are many other examples I could give and arguments I could make, but the essence is this: an artist looks at nature, observes a particular scene, and creates the "explanation" that we call a landscape or a symphony. Nature, however, does not possess "landscapes." Nature presents us with a dynamic scene: birds are flying, leaves are moving, clouds are changing shape, and the light keeps shifting in hue and intensity. From this wealth of information, the artist extracts a static essence, the "signal" from the "noise," and creates a landscape. If we stare long and hard at the landscape and then back at nature's own scene, we too can see the landscape. But it is not really there. It is not nature's creation; it is the artist's.

A scientist also observes nature, some particular phenomenon, and creates an explanation that we call a theory. The explanation is just as much as an abstraction of the real phenomenon as the landscape is of nature's scenery. If the theory takes a sufficient grip on the collective mind of society, it becomes such a natural way to think about things that it actually becomes a discovery. Once Newton convinced us that the apple was being pulled down by a force, we forgot the creative chain of reasoning that produced this explanation. The force became real. In a sense it leapt from Newton's mind and became an action between the earth and the apple. There is an expression that we all use: "I won't believe it until I see it!" I would like to have you think for a moment about the converse of this expression: "I won't see it until I believe it!" What I mean by this is that once we believe in some basic set of concepts, once we have some world view or philosophical infrastructure, we begin to fit things into it that we would never have noticed before. Our beliefs give form to our observations and give them the status of discoveries. Now I have to convince you that Newton's creation should be replaced by Einstein's.

Mathematics and Creativity

The artist's creation must be expressed in some medium, be it paint on a canvas, a block of marble, or the sounds of an orchestra. The scientist's creation is expressed in the symbols and logic of mathematics. Mathematics, with its richness and diversity, provides a medium that, at least up to the present time, has been sufficient to create an acceptable reality. In addition to its richness, the language of mathematics has an extraordinary depth. In contrast, a language like English is very broad but also very shallow. Its breadth lies in its capability of describing anything. There is virtually no subject that cannot, in principle, be discussed in English. Yet English is also shallow, in that one cannot engage in a discussion for very long before ambiguities arise and disagreements occur. Utter enough words on any topic, and whatever you say becomes subject to interpretation. Mathematics, on the other hand, is much narrower than English in terms of its legitimate subject matter. Everyday sorts of things cannot be discussed effectively using calculus or differential geometry. However, when mathematics is applied correctly and legitimately, the "conversation" can go on almost indefinitely with little if any ambiguity or disagreement. This extraordinary depth of mathematics allows and sometimes compels the scientist to exercise a certain type of creativity, but it is limiting as well. We shall have further occasion to discuss these limits as we go on.

The Limits Imposed by Our Minds

In the chapters that follow, we shall begin by discussing physics in general terms. Before examining specific areas, like statistical physics or quantum mechanics, I will explain what contemporary physicists are trying to do and how they are going about it. From the start, doing physics is limited by the nature of our minds. The mind, after all, is the primary tool that we use to investigate all phenomena. If we find that this tool has certain flaws, predispositions, or idiosyncrasies, we may have to adjust for them or at least learn to interpret our results in the light of them.

Most people find it hard to believe that our minds impose fundamental limitations on our ability to explain the behavior of the world around us. We have no difficulty accepting the fact that our bodies impose limitations on our physical abilities. The speeds at which we run, the heights to which we jump, and the weights we can lift are ultimately limited by our physiology. None of us questions or complains about these limits. Although great athletes seem to be constantly extending them farther than we thought possible, the extensions become smaller and smaller and are increasingly difficult to achieve. Similarly, there are limits to the range of colors we can see, the details we can visually resolve, the sounds we can hear, and the odors we can distinguish, limits that are also set by our physiology. We don't have difficulty accepting these limitations either. Why then should we be less willing to accept the existence of limits on the way we think about things? In fact, the most exciting thing about studying contemporary physics is learning how we have come to accept our limitations and arrive at an understanding of nature in spite of them. I like to use the following analogy. Looking at nature with our mental limitations is like looking at a scene with poor eyesight. The scene is fuzzy and distorted. Modern physics provides us with a pair of conceptual eyeglasses that correct our vision and thereby give us a clearer image of the real world.

Three

The Physical and Mental Limitations of Humans

Physical and mental limitations severely affect both the way we observe the world and the way we think about it. Our physical limitations are obvious. Our eyes can see the world around us only because they are sensitive to visible light, which scientists know is an extremely limited band of electromagnetic radiation. The world is filled with other sorts of radiation to which we are totally oblivious. Anyone whose eyes (or other organs) were sensitive to radio waves or infrared or ultraviolet radiation would unquestionably perceive a very different world than the rest of us do. While we are now able to construct devices that are sensitive to these other kinds of radiation (e.g., infrared sensitive film, radios, X-ray film, etc.), our own personal physiology lacks this sensitivity. I might add that there is a very strong indication that, among certain animals, the brain fails to process some signals the eyes send to it. It does not permit the eyes to "see" everything that falls on the retina and is transmitted through the optic nerve. Thus there appears to be further neurological filtering and processing of the already sadly incomplete picture of the world we receive directly from our sensory organs.

Our hearing is restricted to a very narrow range of sonic frequencies. A dog can hear frequencies that are far beyond the human auditory range. Fish are believed capable of sensing frequencies in water that are lower than any we are capable of perceiving. Our sense of touch is limited by the size of our hands, the span of our arms, and the sensitivity of our nerves. We can't grasp a molecule or spread our arms wide enough to embrace a galaxy, and we can barely feel the air that surrounds us. Even when we admit to these physical limitations, however, we still tend to maintain—quite foolishly, perhaps—that we have a good sense of the real world!

Along with the purely physical limitations on our ability to acquire knowledge, we are afflicted with a host of mental restrictions. By this I mean that the brain does not just passively "process" information, but selects and manipulates it according to rules established by hundreds of thousands of years of genetic development. This is a controversial notion, but I will attempt to convince you that it is true. Some clues to the way in which our brain has developed have been given us by workers in the field of paleoneurology. This discipline studies neural development by analyzing the fossilized skulls of ancient and modern animals. The changing shapes of skulls give paleoneurologists information about how the structure of the brain must have changed in time, and since we know what portions of the brain control different aspects of our behavior, we can infer how behavior itself must have changed in time. Indeed, according to the Darwinian theory of evolution, not only did the changing brain affect our behavior, but our changing behavior patterns selected those developments in the brain that made it a more favorable tool for supporting those patterns. Thus one could say that behavior affects the brain just as the brain affects behavior. What then makes one brain more favored than another? Why should nature retain one type of neural development and allow another to die out? The paleoneurologist Harry Jerison, in an essay entitled "Paleoneurology and the Evolution of Mind" (*Scientific American* 234, No. 1 [Jan. 1976]) makes the following statement:

> Reality, or the real world we know intuitively, is a creation of the nervous system; a model of a possible world, which enables the nervous system to handle the enormous amount of information it receives and processes. . . . The "true" or "real" world is specific to a species and is dependent on how the brain of the species works; this is as true for our own world—the world as we know it—as it is for the world of any species. The work of the brain is to create a model of a possible world rather than to record and transmit to the mind a world that is metaphysically true.

I would like to emphasize the words "creation" and "model" in this statement because I will use them in precisely the same way when I discuss physics. If the brain itself "creates" the reality in which we live out our daily existence, it is not surprising that the brains of great thinkers create the more subtle reality that occupies our intellectual attention.

As part of its role in creating our reality, the human brain almost certainly exercises some control over the quality and quantity of sensory data that are transmitted to it. Some neurobiologists claim that this is how our brains insure that they receive only the amount of data they are capable of handling. But there is also a more subtle kind of mental limitation that shapes the nature of our knowledge. This limitation results from the fact that there is a set of conceptions, a way of thinking about things, that is also quite probably built into our brains, either genetically or by learned acquisition. Loosely speaking, I will call this "common sense." There are fancier and more pretentious phrases for it, such as "deep structures" or "cognitive strategies," and I have even heard the phrase "conceptual gridwork," but I don't think these terms necessarily convey any more meaning than old-fashioned "common sense."

We acquire many common-sense conceptions through formal education, and we learn others from day-to-day experience. Still others we acquire through contact with our culture and society. This is all reasonable and understandable. What is quite surprising, however, is the fact that some highly respected scientists believe certain of these concepts are "innate" and may actually be genetic in origin. This means that they have become a part of our brains by virtue of the evolutionary process. We do not learn them; we inherit them. This is a very radical position, suggesting that we are born with some reasoning faculties already in our posession. Our brains are not the "clean slates" ready to be written on that some philosophers once believed they were. Those who have purchased a computer know that when it was removed from its packing crate and turned on for the first time, it already knew how to greet us with a smiling icon and message of welcome. Computers are equipped with an operating system. Our own situation is analogous. We are equipped with genetic knowledge, not what we have learned as individuals, but what we have learned as humans.

If you have studied the Darwinian theory, you know it is driven by "adaptation" and "survival." The body and the brain within it are the product of hundreds of thousands of years of adapting to their environment and surviving to procreate. It is not unreasonable to assume, therefore, that whatever "operating system" presently serves the brain is there primarily because of its survival value. The evolutionary road to modernity is littered with the corpses of humans that were not as well equipped to procreate as are the bodies of the survivors, those you see around you.

I'm not convinced that studying physics has any special survival value, so I can't argue that the brain is particularly well equipped to do that. Nevertheless, we do study physics, and some of us even like to think we do it well! We have to be prepared to accept the fact that we do things with our brains today that, evolutionarily speaking, they were not really designed to do. There are several kinds of concepts we bring to the intellectual task of understanding nature simply because they "make sense," or more likely because they are a part of our genetic operating system.

Three Innate Conceptual Structures

I would like to focus on three concepts that are as important in physics as in our everyday lives, "cause and effect," "reductionism," and "space and time." All of us use these concepts constantly and believe them to be true, yet none of us have learned them at any identifiable time or place. Is that a sufficient reason to claim that they are "hard wired" into our brains? I don't have proof of this, one way or the other; therefore I'm not going to claim that a neurosurgeon could locate some segment of the brain and call it the seat of "cause and effect." Nevertheless the impact of this concept (and the others) on the way we think about our very existence is so important and pervasive that I will claim it to be a genetic development.

What exactly are these three concepts? Cause and effect is simply the notion that anything that happens does so "because" of some particular preceding event. Things don't happen spontaneously or at random, even though it may often seem that way. Rather, life proceeds in a sequence of linked happenings, and the linkages allow the happenings to meaningfully affect each other. Now it might be asked, why would such a concept be handed down genetically? What is its survival value? It is quite possible that there were some early humans whose brains viewed the world "acausally" and saw in the slings and arrows of outrageous fortune neither rhyme nor reason. These humans were all struck by lightning because they never learned to associate big black clouds with thunderstorms and thunderstorms with danger. In short, there are so many causal sequences of events in nature that, if not recognized, would wipe out those possessing acausal brains, that today we find neither remnants of such brains nor of the creatures unfortunate enough to have possessed them.

Reductionism is possibly an even more basic conception than cause and effect: the belief that all complex processes or systems can be understood in terms of their constituent parts. All things can be reduced to the operations of their simplest components. From our earliest days as children we took things apart to understand them. (We couldn't always put them back together, but that's another story.) The opposite of reductionism is "holism," the belief that things are "other than" the sums of their parts. Today we hear quite a bit about holistic medicine, an attempt to treat the body as not simply a collection of organs, but an identifiable entity. The vast majority of doctors, however, are devout reductionists, and virtually all of the successes of modern medicine are reductionist in origin. Holism does not come naturally to most of us, and its applications seem strange. If your car doesn't start some morning, you automatically check off a list of possibilities: Am I out of gas, is the battery dead, has the ignition system failed? This kind of checklist is the earmark of a reductionist. We think of the car in terms of its parts. The comedian Jack Paar once told a story of his father being so angered by his car not starting when he cranked it that he smashed one of its two headlights with the crank handle and then advanced upon the other one, shouting, "Now I'll blind you!" This man was clearly a holist, thinking of the car as a malign entity rather than simply the sum of its parts. Such stories notwithstanding, reductionism is a window through which nearly all of us view the world.

Finally, there is the combined concept of space and time, which form the stage where all the events in our lives take place. It is this dual conception that allows us to entertain the notion that events occur at a certain place and a certain time and, moreover, that these labels are fundamentally different. You find your position in space with a map (or a ruler and compass) and your place in time with a clock. You can move forward, backward, and up and down in space, but only forward in time. Without the conception of space and time built in to your brain, you could not even begin to think about descriptions of the world. All you have to do is look around you to realize that your mind presents you with a visual space or "field." Objects fill that visual field. When you look "outward" at the world around you, you are really looking "inward" at the visual field within your mind. Your brain projects this inner picture outward and provides you with the sense of viewing an external world. But make no mistake; you are really looking at your brain's very own creation, not directly at an independent reality. You create the external world from your own inner sensations. I pointed out in the first chapter that this idea is by no means original with me, noting an artistic representation of it by the Belgian surrealist René Magritte.

Your mind also has a built-in clock that provides a sense of time to accompany its built-in sense of space. You have a definite sense of time moving forward, from a past that has already occurred to a future that is yet to be. Strengthening this conception is the interesting neurological feature called memory. Memories are always of the past and never of the future (with the possible exception of the phenomenon called "precognition," which is amazingly akin to a memory of the future). Why is that? Is it because, as we all "know," the future has not yet occurred? Or is it possible that the future already exists, but there is something that prevents us from knowing about it? Perhaps time, like a one-way road, has no intrinsic direction, but has been caused to appear to have one by other agencies.

However it may be that the conception of space and time comes into being, it is a precondition of rational thought about the external world. That space and time exist and have the properties we know they do is not something we are taught in school; it is a genetic inheritance.

The thrust of this book will be to demonstrate that these three most cherished and useful concepts are, in fact, wrong! We have them, we use them, we trust them, and we apply them to all our intellectual pursuits, but they are fundamentally flawed. I will endeavor to show why they are flawed and what the consequences are of using them with and without their imperfections.

Concepts, Experience, and the New Realities

All three of the concepts I have just discussed were developed as a result of our experiences. Not just our own personal experiences, but also our genetic experience. Biologists use the expression "ontogeny recapitulates phylogeny" to express the fact that the fetal development of a particular individual (ontogeny) is simply a smaller-scale version (recapitulation) of the way the entire species has developed over time (phylogeny). The human fetus passes through stages in which it has a tail, gills, and the like. I make the same claim for conceptual development. Individually and personally, we make use of concepts that the entire race has found useful or necessary for survival.

Our experiences, personal and genetic, have all taken place here on Earth. Moreover, all have been with phenomena involving systems that are of our size and move at the speeds we move at. No human has ever lived on an atom or touched other solar systems. No human has ever been able to outrun a speeding bullet. I have already called the world we inhabit and our experiences within it "macroscopic" reality, because that is the standard textbook term; but I much prefer "mesoscopic," because it is more accurate. As noted in the first chapter, mesoscopic means "middle-sized," and it refers to everything about us—our size, our speed, and so forth. Our brains are mesoscopic. We think mesoscopic thoughts, we dream mesoscopic dreams, we take mesoscopic courses from mesoscopic teachers. It should be of no surprise that cause and effect, reductionism, and space and time, however we acquire them, are mesoscopic concepts.

If mesoscopic reality were the only reality, there would be no problem associated with having a mesoscopic brain. However, as I've already suggested, there are other realities. What happens if we try to imagine life on an atom using our mesoscopic imaginations? In other words, can we legitimately apply cause and effect, reductionism, and space and time in an attempt to describe or understand an atomic or subatomic world? If we do so, we are applying concepts born of our experiences to the understanding of experiences that are completely beyond the plane of our existence. If I could shrink you down to subatomic size and place you on an atom, would your survival require cause and effect, or would it be hindered by it? We shall see.

If it were simply a question of applying our experiences to other realities and testing them to see if they worked, the matter would be simple. Unfortunately, other realities are other realities precisely because we *cannot* live in them. We can think about them and, to some degree, experience them through the effects they have in our reality, but we cannot truly live in them. If we could, they would be part of our reality. As a result, we think about other realities in mesoscopic terms and then test our explanations and predictions back here in our mesoscopic world. The realities that we come to believe in as a result of conceptualizing with our mesoscopic brains and testing these conceptions with mesoscopic experiments are therefore created realities. I have already said that I like to think of created realities as kinds of maps. It is not a question of whether or not they are true, only of whether or not they work. And they have to work for us, in our mesoscopic reality. The Mercator projection is a map we have all learned to recognize. For many of us it *is* the real world. Its value lies in the fact that it works extremely well for navigational purposes. There are many maps that offer a much better model of the world, but they are inconvenient to use.

Discovering errors in created realities is very difficult, and when we do find errors, they are often so strange and paradoxical that it is difficult to figure out exactly what is going on. In a very real sense, once the field of physics went beyond its traditional task of describing macroscopic reality (to return to the textbook term) and attempted to create maps for other realities, it became as much the study of knowledge per se as the study of things. Physics also becomes, in this way, a study of ourselves: how we know things, what the limits are on what we can know, and how we know when we have reached these limits.

From the time of Isaac Newton in the seventeenth century until the middle of the 1800s, the preponderance of research, observation, discovery, and explanation in physics dealt with experiences that were directly within our macroscopic reality. This was the age of "classical" physics, and it was truly a golden age marked by innumerable successes, including Newton's explanation of the motion of material particles (classical mechanics); the description of the thermal behavior of matter (classical thermodynamics) by the nineteenth-century scientists Sadi Carnot, Lord Kelvin, and James Joule; and the incorporation of a host of seemingly disparate electrical and magnetic phenomena under a single, all-encompassing theory (classical electrodynamics) by their contemporaries André-Marie Ampère, Michael Faraday, Carl Friedrich Gauss, and James Maxwell. All areas of classical physics were consistent with cause and effect, reductionism, and the use of space and time for both descriptive and explanatory purposes. In fact, the mathematical superstructure of Newton's mechanics further enlarged and strengthened the concept of cause and effect with an additional notion called "determinism." Determinism is a word that describes the behavior of certain kinds of mathematical equations, whose solutions lead inexorably from some given set of initial conditions to a unique final result. Newton's equations of motion are deterministic. If the position and velocity of Earth in its orbit around the Sun are known precisely at any given instant of time, its position and velocity will be known with similar precision for all succeeding instants of time. I do not wish to get into the philosophical subtleties of distinguishing between cause and effect and determinism, but suffice it to say that the age of classical physics, dating from Newton's monumental works, treated cause and effect and determinism with almost a sense of reverence.

From the 1860s onward, however, as noted in the first chapter, scientists began investigating a wide range of new phenomena that went beyond reality on our usual scale. The question arose as to whether our mesoscopic constructs would retain their validity and usefulness in these new areas. In chapter 1 I called these new areas of investigation "new realities." In the following chapters I will give a brief account of what these five realities encompass and why it is not unreasonable to expect that our mesoscopic brains should be inadequate for the tasks of exploring, mapping, and explaining them. I give you fair warning that you will not be able to understand some of the new concepts. The fault is not your stupidity or poor study habits; it is because you have a genetic disability!

I do not want to give the impression that realities beyond our mesoscopic reality are important only in the study of physics or that they involve ideas of such subtlety and abstractness that they have no relevance whatever to our everyday existence. This is simply not the case. On the contrary, they will cause you to confront presuppositions wherever your own intellectual interests reside. In history, for example, just as in physics, one can think in terms of a microscopic, a mesoscopic, and a macroscopic reality. These realities depend on the time scales with which we view the course of human events, the spatial scales on which we interact with our surroundings, and more subtle scales of comprehension that depend on information and memory. Viewed from the microscopic scale, we might imagine the French Revolution as perceived by a lowly citizen of Paris, let us say a *boulanger* (baker) living during those fateful days. Would this hard-working baker even know that a revolution was taking place? He awakes every morning to the aroma of his good wife's fresh *café*. He wends his way to his *boulangerie* to prepare *les croissants, les tartes aux fruits*, *les pains*, and so forth. Then, when the day is done, he goes home. He measures history on the scale of a day. His comprehension of the events of the day depends on the number of people he meets, the information they bring him, and the degree to which he can assimilate that information. It is quite conceivable that on this scale the revolution, as an event, does not even exist. Perhaps "history," as a sequence of recognizable events, exists only on a mesoscopic scale, involving years and societies as opposed to days and individuals. I can't help thinking of the old colored comic strips in the Sunday papers, which were composed of an enormous number of tiny colored dots, or the magnificent paintings of the French pointillist artist Georges Seurat. If you examined these pictures up close, all you could see were dots; the picture itself disappeared. Is this the analogue of history on the human scale?

Although I will not go any farther at this time, I could just as easily have chosen economics as a discipline with a mesoscopic, microscopic, and macroscopic set of realities. Before leaving this topic, however, let me add one more comment about the concept of history. We all believe that there is really only one history. It is a collection of facts, the course of human events laid out in a fixed and immutable fabric of space and time. Different people may look at this same fabric and interpret it differently, but we are all in agreement that it is the same fabric they are looking at. The course of history is objective; different interpretations of it are subjective. Is this in fact true? Is it not possible that there are as many different histories as there are interpreters, that the collection of different subjective views is all that exists, and that there is no underlying fabric of objective history? We shall see that Einstein replaces the Newtonian concept of a single absolute space *and* time with an infinity of separate spacetimes (note the absence of the "and"), each one belonging to a different observer. I shall, in fact, spend a great deal of time explaining how and why Einstein did this and what its implications are for physics and the more general way we view the world around us. But can we extend Einstein's approach to history as well?

How do we even know that a cosmic reality exists, or, if it does, that it is fundamentally different from the macroscopic scale? This is a very good question, and I will soon answer it in great detail. But let me whet your appetite by giving you a glimpse of the cosmic world that can be obtained from within the confines of our merely macroscopic world, like the dragons and sea serpents that ancient mapmakers drew along the boundaries of their maps to mask their ignorance of the seas beyond. Imagine yourself out some evening with a particularly close friend. You stare upward at the vast open spaces of the evening sky and gaze at the myriad twinkling stars that spread themselves across the void. "Ah," you say, "just look at those stars; isn't space enthralling!" It may very well be, but space isn't what you are looking at. You are, in fact, looking at *time*! None of the stars that have you so enraptured are even there, where you now see them. Every twinkling dot is light that was sent out by that star millions or billions of years ago. What is up there tonight, or should I say what your brain projects outward as being "up there," is a vast illusion, a collage of space and time, made up of light from objects that are nowhere near where they appear to be. Illusions such as this require vast distances and times in order to come into being. They belong to cosmic reality. Such illusions do not exist in microscopic reality, but others equally interesting do. It is just one example of how we can be fooled when we intrude on another reality with a mind made for this reality.

Four

Symmetry

Scientists often impose their own aesthetic predispositions on nature through the theories they devise to explain it. Aesthetic concepts are often used to choose between competing ideas or theories that otherwise might be equally compelling. "I prefer this theory because it is simpler" is an aesthetic choice that is not different from "I prefer this piece of music because I can whistle the melody." Einstein once said, "The Lord is subtle, but he is not malicious." This expressed his belief that nature could ultimately be understood, although the process of learning would be challenging. As it turns out, Einstein's theories were remarkable for their subtlety and simplicity. Does this mean that, in fact, nature has those characteristics, or does it mean that Einstein created a nature with those characteristics?

In science, one of the most universal of all aesthetic concepts is the concept of symmetry. Although related to the highly subjective idea of "beauty," symmetry can be expressed in thoroughly objective mathematical terms. To the scientist, its mathematical formulation makes symmetry one of the most useful of all the aesthetic attributes. We will ultimately see how symmetry has been applied to produce some of our deepest theoretical insights into nature, but we will begin simply, by defining it and discussing how symmetry applies to geometrical objects.

An object is said to "have a symmetry" (or, in more common parlance, to "be symmetrical") if there is something that can be done to it that, as far as we can tell, leaves it unchanged. This particular "something," this undetectable operation, is called the symmetry operation or transformation. The "as far as we can tell" part may mean that we simply inspect the object visually and fail to see any differences in its appearance, or it may involve the application of more subtle and complex types of experimental measurements. The symmetry operation usually has to be some real action, something you can actually do to the object and even tell your friends to do. Certain symmetry operations (some examples will follow) are rather unusual, and it might not be physically possible to do them. These are called "abstract" operations and they will have to be done in your mind or with pencil and paper.

The best example of a symmetry operation comes from a class called geometrical symmetries. These are undetectable transformations of real physical objects that are accomplished by moving the objects in specific ways. A square piece of paper uniformly colored on both sides has several such geometrical symmetries. Suppose I hold the square so that one of its sides is facing you. I then ask you to close your eyes, and, while they are closed, I rotate the square through certain angles (90°, 180°, 270°, etc.) about an axis passing perpendicularly through its center. When you subsequently open your eyes, you will not be able to tell what I did to that square, or even if I did anything at all. I could repeat the experiment and this time turn the square front to back. The same result occurs. You simply can't tell what's been done. That's why these are symmetry operations (or just plain "symmetries").

A rotation of 45° or 30°, however, is not a symmetry. After you opened your eyes, you would instantly see a rotated square. Neither is a translation, a shift of the paper in some arbitrary direction.

Breaking a Symmetry

If I place a visible mark on one side of the paper at its center, I will destroy the paper's front-to-back symmetry, but not its other symmetries. If I put the mark off center, I will destroy many of the other symmetries as well. Rotating the paper through any angle that is not a multiple of 360° will also cause the mark to rotate, and the movement of the mark will be easily seen. Using a mark of some sort to destroy a symmetry is an example of what is called a "symmetry-breaking" device. In Western aesthetics, symmetry is usually considered to be an attribute of beauty. Because of this particular prejudice, we call marks that destroy symmetry "blemishes." In some Eastern cultures, it is the breaking of a symmetry that is considered beautiful. In such cultures, a broken symmetry might be called a "beauty mark." In the realm of simple geometric objects, a piece of paper cut into a circle has many more symmetries than a square. A spherical object, of course, is about as symmetrical as you can get insofar as rotations are concerned.

The collection of all the things you can do to an object that leave it unchanged—that is, all of its symmetry transformations—form what mathematicians call a "group." A group is a set of operations with the following properties. The successive application of any two operations, called group multiplication, must also be an operation that is in the group. One of the operations must be the "identity," an operation that does nothing. Every operation must have an "inverse," an operation that undoes the original one. Clearly, the symmetries of a geometrical object have these properties. If you follow one symmetry transformation by another, the object is still unchanged. The operation of "doing nothing" actually exists and is a legitimate symmetry. Finally, every rotation can be performed in the opposite sense, which undoes it. The study of symmetries is thus rewarding to mathematicians and physicists alike.

Some Unusual Symmetries

The human body posesses a rather strange kind of geometrical symmetry called "bilateral" symmetry. Suppose I take a perfect mirror and cut out of it a silhouette of my body. Let me then step into that mirror and position my body half-way within the cutout. If you then stand in front of the mirror and look at me, you will see a combination of the half of me that remains outside the mirror and its reflection in the mirror. The other half of my body, of course, is hidden behind the mirror. An observer who is unaware of the presence of the mirror, looking at this combination of me and my reflection, would, in principle, be unable to distinguish it from from my entire body. Of course our bodies have so many blemishes that no human could possibly pass this symmetry test. Certainly our internal organs do not have this symmetry, so an X-ray machine would not be fooled.

Ideally, one tries to discover a symmetry by actually doing something to an object and then inspecting it visually or making appropriate measurements of it, to see whether the effects of that "something" can be detected. Depending on the nature of the object, this approach may not always be practical. If the object is a mountain, for example, it would be impossible to rotate it in order to verify that it has a rotational symmetry. One way of avoiding the problem of rotating bulky things like mountains is to leave the mountain in place but rotate the observer. What is important here is the relative movement of mountain and observer. From this point of view, symmetries can be investigated either by moving objects or by moving observers. Let me make one rather subtle point. It is not inconceivable that a case might arise in which these two approaches are not interchangeable. In that event, it would be the object's movement that takes precedence.

The Mathematical Representation of Symmetry Operations

By moving the observer rather than the object, we can investigate many more symmetries than would otherwise be possible. Nevertheless, many symmetries are still extremely difficult to verify by transforming either the object or the observer. An enormous number of symmetry operations are abstract; they can be extremely difficult, if not impossible, to perform directly on a real object. The analogous process of moving the observer is also difficult to conceive in the case of an abstract operation, unless one wishes to use the observer's imagination to allow the process to occur. The bilateral symmetry of the human body discussed above required a somewhat abstract operation for its verification. Perfect mirrors are hard to come by, and cutting perfect silhouettes out of them is equally difficult. Consequently, I relied on your imagination.

A much better way to perform a symmetry operation is to devise a method that does not require dealing with either the real object or the real observer. This can in fact be done through the use of mathematics. I will now show you how to represent an object and its observer in mathematical terms and how to then transform the representation of the observer. Instead of rotating a circular piece of paper relative to an observer, we will represent it by the equation of a circle and then "rotate" the equation.

An equation represents a curve or a shape by providing a relationship between the coordinates of all the points that lie on that curve or shape. If the points lie on the curve, they must satisfy the equation; likewise, if solving the equation produces a certain point, it must lie on the curve. In a rather strange but very real sense, the equation is a shorthand, ultracompact version of the curve itself. It contains all the information about the curve that a drawing of the curve would contain. You can write down the equation of any given curve, even a curve that is infinitely long, and pack it up and take it with you. Open it up at any later time, at any other place, and you can reproduce the curve. Using an equation requires a coordinate system in which to locate arbitrary points. To keep things as simple as possible, we imagine a curve in only two dimensions, which can be represented in a coordinate system consisting of an x and a y axis. If the curve is a circle of radius R, it can be represented by equation (1):

(1) $$x^2 + y^2 = R^2.$$

You should think of the coordinate system as actually *being* the observer. When I write "x, y," think of the observer looking at a certain point that is x inches (or feet or meters) away in a horizontal direction and y inches away in a vertical direction. The observer's head (or brain, or mind, or eyes, or visual field, however you want to think about it) is fixed at the origin (0, 0) of the coordinate system. The equation of the circle, therefore, represents what the observer sees when looking at every point whose x and y coordinates satisfy equation (1). Imagine, in short, that the equation actually *is* an observed circle.

Suppose I wanted to move the circle 5 inches to the right in the horizontal direction (parallel to the x axis). In terms of the coordinate system, this means moving the center of the circle from (0, 0) to (5, 0). Alternatively, I could move the observer 5 inches to the left. This means moving the coordinate system to a new position, with its origin 5 inches to the left of its original position. What would the observer see after being moved 5 inches to the left? In mathematical terms, I am asking what happens to the equation of the circle when I move the coordinate system in which it is described, because it is the equation that represents what the observer sees. This kind of coordinate shift is relatively easy to do. To avoid confusion, we will use the new symbols x' and y' to represent the axes of the moved coordinate system, which also represent the visual frame of the moved observer. We will now express the equation of the circle in these new coordinates and compare it with the equation in the old coordinates. The "transformation of coordinates" written in (2) is the mathematical relationship between the old and new systems:

(2) $$x' = x + 5;$$
$$y' = y.$$

The two equations in (2) express the fact that the original origin, (0, 0) in terms of x and y axes, has now become the shifted point (5, 0) in terms of the x' and y' axes. It is very important to remember that whenever an observer (or an object) is moved or transformed, you must introduce new labels for the coordinates so as not to confuse what the moved observer sees with what the original observer saw.

Using (2) to transform equation (1) is quite simple. Just replace x and y in the original equation with $x' - 5$ and y'. This gives

(3) $$(x' - 5)^2 + y'^2 = R^2.$$

The transformed equation (3) does not have the same form in x' and y' that it had in x and y in equation (1). This is the crucial result. Because of the new form of the equation, we can say that the observer no longer sees a circle that is directly in front but rather 5 inches to the right. "But we already knew that," you say. Of course we did, but we have to make sure that the mathematics reflects what we know. Seeing a different circle is the same thing as the new circle having a different equation. This is the key to examining symmetries from the mathematical perspective. If your eyes can't detect any differences, the equation has to look the same after the change in coordinates as it did before. Instead of asking, "Does the object look the same to the moved observer?" we will now ask, "Does the equation have the same mathematical form in the new coordinate system?"

When we evaluate symmetries in terms of the behavior of equations, we often call them "invariances" instead of symmetries. The object has or doesn't have a symmetry; the equation has or doesn't have an invariance. This shift, from examining the appearance of an object to examining the form of an equation, will give us amazing flexibility in discovering new symmetries. This is particularly true because the laws of nature, which we will be examining for symmetries, are already given as equations. There are no objects to examine.

Now that I've demonstrated that moving an observer 5 inches to the left of a circle is a clearly detectable operation, I would also like to demonstrate that rotating a circle clockwise through some arbitrary angle α is undetectable, and therefore a legitimate symmetry operation. Unfortunately, arbitrary rotations are not easy to describe mathematically, so I will pick a particular angle, $\alpha = 90°$. Since a clockwise rotation of the circle is equivalent to a counterclockwise rotation of the observer, we will need the mathematical transformation corresponding to a counterclockwise rotation of our coordinate axes. Therefore, I need to know how to rotate the x and y axes counterclockwise through an angle. For this particular choice of $\alpha = 90°$, the new y axis (called y') is the old x axis and the new x axis (called x') is the old negative y axis. In the language of transformations,

(4)
$$x' = -y;$$
$$y' = x.$$

Inserting (4) directly into (1) gives:

(5)
$$x'^2 + y'^2 = R^2.$$

This, in the primed coordinates, is the same circle equation that we had in the unprimed coordinates. Thus we can say that the rotated and unrotated observer see the same circle.

Symmetries of Nature

I would like to progress from the symmetries of physical objects to a broader but less obvious concept, the symmetries of nature. Nature is a complex collection of objects and processes that involve objects. By a symmetry of nature, therefore, I mean something that can be done to some set of processes—possibly all of the processes that occur in nature—yet will be undetectable to an observer. Alternatively, I also mean anything that can be done to an observer that leaves the observer's description of the processes unchanged. For example, here on Earth you observe that all things released from rest fall down. If I could somehow turn you upside down, you would see things falling up. Therefore, an "inversion" of an observer is not a symmetry is the case of falling objects on Earth.

This is a rather over-simplified approach to how symmetries in nature are discovered. In general, a transformation has to be applied to an entire set of naturally occurring phenomena or to an observer of those phenomena, and the phenomena, taken as a group, must appear the same after the transformation is applied. Let me stress an important point here. It is not necessary that each individual phenomenon appear unchanged for a symmetry to be declared. It is sufficient if the set of phenomena remains unchanged. Phenomenon A might turn out looking just like phenomenon B, but then one of the other phenomena must wind up looking like A. In this way, it is the collection that cannot be distinguished after the symmetry operation, not its individual members.

In discussing symmetries of nature, I will continue to speak of the observer being transformed, not the observed phenomena. I will also continue to use completely mathematical descriptions, although I will discuss the meanings of symmetries from the standpoint of a physical observer as well.

In the mathematical description of nature (a field of study associated mainly with Newton) the human observer is in essence replaced by a set of three spatial coordinates (x, y, z), which represent the position of a phenomenon relative to the observer, and one time coordinate (t), which represents the time when the phenomenon occurs relative to the observer. We are doing exactly what we did before when we translated and rotated observers viewing circles, only now we are going to translate, rotate, and do a host of other things to observers viewing natural phenomena. It is much easier to manipulate coordinates than observers, and it is much easier to manipulate observers than phenomena. One of the complaints that I sometimes hear voiced about science is that it is "dehumanizing." Certainly telling someone that they can be replaced by a coordinate system seems to take a big chunk of humanity out of scientific description. Yet, as we shall see, the ability to deal as much as possible with coordinate variables and as little as possible with human beings is precisely what gives science its strength in developing an understanding of certain phenomena.

The final portion of this introduction to symmetry concerns the more general question Why? Why should symmetry be of interest, aesthetic or otherwise, to a scientist? I have already noted that concepts such as cause and effect, reductionism, and space and time have come out of our experiences with macroscopic reality. We therefore have no reason to believe they will work in the other five realities. In fact, each, to a certain degree, doesn't. Symmetry, however, seems to permeate all the realities. Even though we arrived at its usefulness macroscopically, it seems to be pertinent at all levels. Perhaps this is because the symmetries of nature, like the symmetries of physical objects, represent some deep level of structure. After all, you can tell that a ball is a ball just by running your hand over it. What you feel is its rotational symmetry. You certainly can't tell if it's a red ball or a white ball or if it is wood or plastic. Its shape seems to have a more fundamental importance than its color or substance. Perhaps this relative importance of certain qualities is true at a more general level of nature as well.

Allow me to suggest another reason that symmetries might be particularly important. Without meaning to appear sacrilegious, I like to think of symmetries as representing aspects of nature that its Creator didn't care about. If the Creator cared about something, I feel sure there would be some mechanism for letting us know if it had changed. But a symmetry is precisely something that you can change without knowing it has changed. Thus, in my opinion, in a certain sense the quantity being changed is unimportant. Symmetries are principles of indifference. Consider this. A potter who crafts a completely symmetric vase is stating, in effect, a lack of concern over how you hold it, use it, or display it. Those considerations were just not important to the potter. The potter who incorporates a handle in the vase is saying, "This is where I want you to hold it." The potter who paints a design on one side of the vase is saying, "This is what I want you to look at when you hold it." The subversion of symmetries, when done by human agency, usually implies a motive, either utilitarian or ornamental.

The statement "All men are created equal" is the declaration of a symmetry. It is not a geometric symmetry, to be determined visually; it is a political/legal symmetry. The symmetry operation is the interchange of "men." The law (i.e., the legal system) should not be sensitive to the replacement of one person by another. If the authors of the Declaration of Independence had cared about the legal significance of differences between particular individuals, they would not have incorporated that particular symmetry into their statement of principles. Of course, just because the symmetry has been built into our system of government, it doesn't necessarily follow that it is always operative. The symmetry can be broken in at least two ways. New laws can be written that do not incorporate the symmetry, or laws that are in fact symmetric can be applied in an unsymmetric manner. Every time a new law is written, it should be tested for that symmetry. Every time the legal system operates, it should be tested for symmetry in its application. We should test laws and systems every so often, by replacing, for example, a man with a woman, or a white male with a black male, or one ethnic group with another. Do the laws operate with no detectable differences? One way to make certain that a new law retains the symmetry is to write it so that it makes no reference to any particular class of persons. This is equivalent to constructing an equation in which a certain variable simply doesn't appear. Clearly, then, changing that variable can have no effect, since there is no place to put the changed variable. In cases where symmetries are produced by simply eliminating a variable from the equations, we may also produce a simpler-looking theory. In those cases, then, symmetry can lead to simplicity.

If we intentionally put symmetries into the theories we are creating, they express our indifference to some features of the theories. Nature's inherent symmetries also represent some important indifference. Therefore, when physicists search for symmetries of nature, they are already expressing an intellectual predisposition toward the existence of indifferences in nature, which may or may not be present. The great physicist and Nobel laureate P.A.M. Dirac was once discussing a particular invariance under a type of "reflection" (called "parity") that had been found to exist in the laws of nature. He remarked, "I do not believe there is any need for physical laws to be invariant under these reflections, although all the exact laws of nature so far found do have this invariance" ("Forms of Relativistic Dynamics," *Review of Modern Physics* 21 [1949], 392–399). It is revealing that Dirac commented on the "need" of this invariance, an expression of his belief that nature should not require any excess baggage in the form of this particular symmetry. To Dirac, therefore, the presence of a symmetry is not some mere chance occurrence but is rather a necessity dictated by some overriding property of nature. What makes Dirac's comment all the more remarkable (and a tribute to his genius) is that less than ten years later it was discovered that nature, in fact, did not always possess this symmetry. There was an unusual class of phenomena that did not have it. As Dirac correctly suspected, the laws "so far found" did have the symmetry, but subsequent laws did not. The loss of this symmetry led to equations that were not as simple, but to Dirac it was more important that nature not allow us to impose our own aesthetics on it. Even Einstein is supposed to have said that "physics should be made as simple as possible, but no simpler!"

What, then, are the symmetries of nature? In fact, there seem to be many. I will discuss some of them, not necessarily in the order of their importance but in the order of their conceptual and technical simplicity. The first and simplest symmetry has come to be called the "homogeneity of space." This symmetry states that an observer can be moved to arbitrary points in space and will be unable to detect that such movement has occurred. In short, space is the same at all points. No experiment you could possibly carry out would allow you to discover that you were at some particular position in space as opposed to any other. In fact, this symmetry can quite easily be demonstrated. Even as you read this paragraph, Earth is hurtling through space, and when you finally put this book down you will be many thousands of miles away from where you were when you began to read. Have you noticed anything different? Have any fundamental physical phenomena changed during this time? Indeed, have any fundamental phenomena changed during your entire lifetime? Let me hasten to point out that this symmetry is beset by many blemishes. The presence of matter in space actually does make different places in space different. In fact, if the symmetry were precisely correct, the very phrase, "different places in space" would be rendered meaningless. The symmetry should be thought of as applying in regions relatively free of matter, which fortunately are the regions that Earth has been moving through during recorded time.

Another symmetry of space is called "isotropy." Homogeneity means that space is the same at all points; isotropy means that it is the same in all directions. In principle an observer could orient a given experiment in any direction in space, and the outcome of that experiment would be unaffected. Again, watch out for blemishes. Certainly on Earth gravity plays havoc with isotropy. We must once again imagine ourselves in reasonably empty space in order to show that this symmetry is valid.

A third symmetry is called the homogeneity of time. This means that the particular time in which an experiment is done is unimportant and cannot be determined. Once again, history tells us that no changes in physical law can be discerned during the entire passage of recorded time. That is really all we have as evidence, but there is every reason to believe that time is completely homogeneous. There is also an isotropy of time, a symmetry between past and future. This symmetry, however, has many subtleties, which we shall shortly discuss in greater detail.

If we truly believe these symmetries, we must make absolutely certain that our theories of nature incorporate them or at least do not violate them. One way to do this is to eliminate spatial position, spatial direction, and temporal position from all theories. If a theory makes no reference at all to position in space, then it will be a symmetry.

Now a reader with some background in the sciences will immediately react to this statement: "What nonsense! All physical theories contain position variables and time variables in their mathematical representations." This is an interesting point. It is certainly true that physical theories are couched in terms of space and time coordinates, but these are "relative," not "absolute," coordinates. A relative coordinate is one that is measured relative to some agreed-on place in your laboratory. The coordinate of your laboratory in space, however, is not known. If I proposed a theory that required locating phenomena with respect to some precise position fixed in space itself, this location would be called an "absolute" coordinate. Homogeneity of space simply means that we can get away with using only relative coordinates; absolute coordinates are never required. Thus, if you should see some physical equation containing the coordinate "x," for example, this is really supposed to be "x-X," where X is the location of the reference point inside the laboratory. The same explanation holds when the time coordinate, "t," is used in a theory. This really means the time measured relative to some agreed-upon starting time for the experimental observation to begin. It is not meant to be an absolute time referring to the birth of the universe or the like.

Homogeneity and isotropy are reasonably straightforward symmetries, and, not surprisingly, they are somewhat less than exciting to contemplate. One that's a bit more interesting is called "parity," the one Dirac discussed, and it is somewhat more abstract. Suppose I ask you to close your eyes and, while they are closed, I place a perfect mirror in front of you. When you open your eyes, your entire view of reality has become a reflected view. My question is simple. Can you tell? Is there anything at all about the behavior of natural phenomena that would immediately say to you, "This is all a reflection"? The answer to this question was always thought to be "No."

Let me interject a word of caution at this point. It is true that the image of some particular phenomenon might look different in a mirror, but the criterion for a symmetry of nature is that the collection of *all* transformed phenomena be indistinguishable from the collection of all untransformed phenomena. For example, if your friend Charlie was right-handed, he would appear left-handed in the mirror. "Aha," you would say, "I know I am looking at a mirror image because Charlie isn't left-handed." But that is not the test. The world of natural phenomena is not made up of your personal friends. Suppose, instead, you were looking in the mirror at an arbitrary collection of people, none of whom you knew personally. Then, all you would see would be a group that consisted of so many left-handed people and so many right-handed people. You would have no way of knowing if you were seeing them "directly," or "in reflection." The point is that the reflected view is a completely possible situation. That's what I mean by a symmetry of nature. Now if it turned out, for example, that all humans were right-handed and, for some reason, could be only right-handed, then the mirror image would show a group composed only of left-handers, a physically impossible situation. In that event, you would be completely justified in asserting that you were looking in a mirror. In fact, if all humans were physiologically restricted to right-handedness, they would constitute a symmetry-breaking device as far as parity is concerned.

Although I have used human physiology to illustrate how you might verify the existence of a parity symmetry, this is a somewhat dangerous approach. Humans are so complex that their appearance or behavior under various types of symmetry transformations may not be a good indication of the truth or falsity of the symmetry. For example, all humans have the larger part of their hearts on the left sides of their bodies. Does this mean that an X ray would provide a symmetry-breaking device for parity? Not necessarily, because it is not clear that the shape of the human heart is really an indication that reflection is not a fundamental symmetry, or if it is an indication of some genetic or developmental idiosyncrasy. Maybe people with hearts that are larger on their right sides are equally possible, but are ruled out by some physiological process that has nothing to do with left versus right. In fact, most of our organs are either doubled or uniformly centered in our bodies, so there might be something special about the heart that is biological rather than physical. It is for reasons like this that physicists limit the test of parity symmetry to purely physical processes, like those involving basic interactions among particles.

Using the criterion of simple, purely physical processes to guide them, physicists arrived at a belief that there were no natural phenomena that would seem different when viewed through a mirror from the way they would look when observed normally. Translated into mathematical language, this meant that the equations used in physical theories would have to be invariant under what is termed "reflection of coordinates." Simply stated, if the three spatial coordinates, x, y, and z, are replaced everywhere they appear by the new coordinates $x' = -x$, $y' = -y$ and $z' = -z$, then every equation would have exactly the same form in the new coordinates that it had in the old. Replacement of the coordinates in this fashion corresponds to having the observer view the world through a mirror.

Now this seems a bit puzzling for a moment. A slightly different way of looking at inverted and noninverted coordinate systems is to form one of each kind using the thumb, index, and middle fingers of each hand. The standard x, y, and z system that we use is called a "right-handed" coordinate system, because it is just like the arrangement formed by those three fingers of your right hand. Point your right middle finger at yourself and your index finger straight up; then the middle finger is the x axis, your thumb is the y axis, and your index finger is the z axis. Your left hand corresponds to the reflection of your right hand (touch middle fingers tip to tip to see this) and also to a complete inversion. To see this, point your left thumb down and your left index finger at yourself (rather awkward), so that your left middle finger points to the right. Now, with your right hand, point your thumb up and your index finger away from yourself, and your middle finger will point to the left. Each finger is the "negative" of its counterpart. Physicists usually refer to the parity symmetry as the equivalence of right- and left-handed coordinate systems. You could phrase the symmetry this way: there should be no difference in the equations of a physical theory if they are described in either a right-handed or a left-handed coordinate system.

In 1957 a rather earth-shattering experiment demonstrated that nature does not possess the symmetry of parity. It was as though a magnitude-7 earthquake had struck the theoretical-physics landscape! Parity, after all, seemed such a reasonable symmetry; why should nature be unreasonable? Thought of in terms of a principle of indifference, why should anyone care whether nature is described by left-handed or right-handed coordinate systems? Remember, however, the remarkable comment by Dirac, which suggested that unneeded symmetries shouldn't be imposed on nature. Apparently nature agreed with Dirac in this case.

The violation of the symmetry of parity was discovered in an extremely subtle physical phenomenon that occurred on the microscopic level of elementary-particle interactions. This is not surprising, since more easily observed phenomena had never even hinted that the symmetry was incorrect. It is not my purpose in this book to get overly involved in elementary-particle physics, so I will comment only briefly on the experiment.

Think of natural phenomena as being classified in terms of four different scales, cosmic (typified by solar systems or galaxies), material (typified by solids and liquids, or even individual molecules), atomic or nuclear (typified by the nuclei of individual atoms), and elementary-particle (typified by protons and neutrons). Careful investigation has revealed that there is a particular "glue" that holds each of these regimes together. That glue is actually a force, and it acts among the characteristic elements of these regimes. It is remarkable that there appear to be only four such fundamental forces in all, and they suffice to keep the entire universe intact. Nature is very economical! Solar systems and galaxies are held together by the "force" of gravity. (I put force in quotes here, because a bit later I will challenge that notion; but for now, "force" is a perfectly reasonable term.) Ordinary matter, as it is found on our own scale (the macroscopic), is held together by the electromagnetic force. Atomic nuclei are held together by what is called (with a physicist's flair for the undramatic) the "strong" force. Finally, certain elementary particles are held together by what is called the "weak" force. I should note that the terms "strong" and "weak" were originally coined to make comparisons with the electromagnetic force, and nobody has come up with better or more acceptable terms.

The motion of objects on a cosmic scale has always been consistent with the symmetry of parity. Any behavior of matter that involves the electromagnetic force was also found to be consistent with parity. Even the processes that involve the strong force and occur on a nuclear level were found to be consistent with the symmetry of parity. Up until the early 1950s, however, the weak force had not undergone extensive scrutiny. This was not surprising, since the individual particle reactions that are affected by the weak force are very difficult to deal with experimentally.

One such reaction was called nuclear beta-decay. A neutron inside an atomic nucleus spontaneously burst apart, to be replaced by three new particles—a proton, an electron, and an antineutrino. It was this process that was found to violate the parity symmetry, in the strangest way imaginable. The antineutrino, like virtually all elementary particles, "spins." Think of it as being something like the planet Earth, which rotates about a north-south polar axis. This spin is assigned a vector sense and direction. The spin vector lies along the axis of spin, and its arrowhead points in the direction your thumb would point if you wrapped your right hand around the antineutrino with your fingers encircling it in the sense of its rotation. The antineutrino is exceptional, however, in that when it moves (always at the speed of light), the direction of its motion (i.e., its velocity or momentum vector) is always along its axis of spin but pointing opposite to the spin vector.

The projection of spin along momentum is called the "helicity" of a particle. The antineutrino's helicity is always negative. Therefore, if you look at an antineutrino coming straight at you, its spin is always clockwise. Counterclockwise antineutrinos simply do not exist. In fact, the unvarying nature of the antineutrino's helicity can be used as an argument that its speed must be the speed of light. If the antineutrino's speed were less than the speed of light, it would be possible for an observer to move faster than it, in which case the antineutrino's momentum would point in the opposite direction to that observer, although its spin would not change. In short, if you could move past an antineutrino, it would appear to have a positive helicity, which is strictly forbidden. The antineutrino's fixed helicity also will change if you look at it in a mirror. In that case its momentum will not change, but the direction of its spin will. Since it is impossible to have a counterclockwise spinning antineutrino (just as it would be impossible to have a left-handed person in a world of all right-handed humans), the mirror image will not represent a possible physical phenomenon and will be immediately recognized as reflected rather than real.

The discovery that antineutrinos have a "handedness" shocked the physics community. It won the Nobel Prize for the two physicists that predicted it in 1956 (T. D. Lee and C. N. Yang) although, strangely, not for the physicist who demonstrated it experimentally in 1957 (C. S. Wu). It is worth noting that all three of the physicists were of Chinese origin. Yang himself speculated that the deep commitment of the Western mind to the bond between symmetry and beauty may have made it easier for an Eastern physicist to conceive that parity might not be a symmetry.

I would be remiss not to mention here one more symmetry that I will discuss later at somewhat greater length. Just as the nonsymmetry of parity can be expressed as our ability to detect the presence of a mirror, so also can the symmetry of time reversal be thought of in terms of a mirror, a mirror that reflects in time rather than space. Imagine looking into such a mirror and seeing in it the universe running backward in time. The question then becomes, Can you detect either the mirror or the fact that time is going in reverse? Our intuitive answer is an emphatic "Yes." All of our experiences tell us that in the real world things run in one direction only. Think how ludicrous the movie of a diver executing a perfect dive into a pool would appear if it were run backward. The diver would seem to be forcefully ejected from the pool by the concerted action of millions of drops of water falling into the pool from above. Then this miraculously launched athlete would gracefully fly heavenward, finally landing perfectly on the edge of a diving board. Is there anyone who would find this a believable scenario? Aside from such particularly humorous situations, the entire process of life is one of aging and death, not "youthening" and rebirth. Only in horror movies do people walk past cemeteries and see corpses emerge and spring back to life. Even something as completely mechanical and simple as a game of billiards would never present us with all the balls bouncing madly from the cushions and then neatly rearranging themselves in a perfect rack. We all feel confident in asserting our ability to detect backward-running movies or time-reversing mirrors.

In fact, time reversal was only quite recently discovered to be a nonsymmetry. The common macroscopic experiences of daily life notwithstanding, every simple microscopic process seemed to look the same in a time-reversing mirror as it did under direct observation. "But what of the more complex processes we have just been discussing?" you ask. While it is true that many such processes looked highly unlikely, almost to the point of laughability, none of them are actually impossible. They are only highly improbable. When we come to discuss statistical mechanics, we shall see that it is possible to construct a theory of macroscopic processes that do not have the symmetry of time reversal, which is based on underlying microscopic processes that are symmetric in this respect. While this theory seems to work quite well and has become completely accepted by practicing physicists, it still seems to present some questions that have not been answered as well as everyone would like.

Other Symmetries

The study of new realities has often revealed new symmetries. It was found, for example, that all subatomic particles possess "exchange symmetry." If any two of the same type of subatomic particles are interchanged, the new arrangement is completely indistinguishable from the original. Any two electrons in a given atom can be exchanged without any properties or characteristics of the atom being modified in a measurable way. Exchange symmetry is absolute, because it implies that all particles of the same type are identical—not identical in the sense of identical twins, which merely look the same, but identical in so strong a sense that they must be regarded as the same particle.

For reasons that are not really understood, all subatomic particles are divided into two classes according to the way in which their identicalness is represented mathematically. One class is called fermions (shorthand for Fermi-Dirac particles); the other class is called bosons (shorthand for Bose-Einstein particles). The identicalness of these particles shows up in their behavior in very unusual ways. A particularly important new form of symmetry that is being investigated, called "supersymmetry," involves the question of what would happen to nature if bosons and fermions were interchanged.

Another important type of subatomic symmetry does not involve changes of the system within ordinary space and time but introduces an "internal" space whose structure has nothing to do with the external world. The members of certain classes of subatomic particles may appear to be quite different superficially, but at the deeper level of this internal space, they are identical. Some symmetry-breaking mechanism (a blemish) in nature produces the seeming difference. Protons and neutrons, for example, are simply different guises of a single particle called the nucleon, but the proton has been dressed up (so to speak) with a positive charge and become differentiated from the neutron. This is a sort of subatomic version of the "nature versus nurture" controversy, wherein identical twins brought up in different circumstances turn out quite differently.

A symmetry in nature must be reflected in the equations that describe nature. The equations must be invariant under the symmetry operation. By the same token, if there is something that breaks the symmetry, it may appear in the equation as a term that does not possess the invariance of the rest of the equation.

Sometimes the underlying invariance of the equation does not show up in each of its separate solutions. The equation may have a set of solutions that, taken as a group, display the symmetry but taken individually do not appear to be symmetric. Thus, reflection symmetry in fetal development could lead to only ambidextrous humans (a symmetric result of a symmetric process), or it could lead to both left-handed and right-handed humans being born in equal numbers (a symmetric pair of solutions). If something having nothing directly to do with the development process makes one of the handedness types appear more often than the other, we say that the symmetry has been "accidentally" broken. A virus might appear that preferentially attacks one type of handedness, for example.

Global and Local Symmetries

All of the symmetries we have discussed so far have been what are called "global" symmetries. The symmetry operation is applied uniformly to all of nature. Thus a single observer may be rotated through some arbitrary angle and asked if everything appears the same. An affirmative answer establishes a global rotational symmetry. We could, however, station many observers at different places in space, rotate each one by a different amount, and see if the resulting composite description (when pieced together) remains unchanged. If the answer is yes, we have a "local" rotational symmetry. If there is a local symmetry, the equations must maintain their invariance under changes that depend on place and/or time.

Sometimes there is reason to believe that a certain local symmetry should exist, even though the equations do not seem to possess it. In such a case, physicists have often "cheated" and added terms to the equation just to produce the symmetry. An example is a "gauge" symmetry, which refers to a method of measuring something, just as a pressure gauge can measure the air pressure in an automobile tire. Since all automobile tires are at a pressure that includes the external air pressure as well as the additional pressure that is applied to the tire with an air pump, gauge manufacturers intentionally make their gauges read zero when they are subject only to external air pressure. The reading is therefore called gauge pressure, meaning that it is the total pressure minus the external air pressure. If something happened to change the external air pressure by a constant amount all around the globe, gauge manufacturers would simply recalibrate their gauges to subtract this new amount. Such a universal change would be called a "global gauge transformation," and it would leave all physical measurements of the important quantity, the tire pressure, unchanged. By analogy, constant changes in certain physical quantities (having nothing to do with air pressure) that do not affect the result of physical measurements are also called global gauge transformations.

Suppose something happened that affected the external air pressure differently in different parts of the world. Then gauge manufacturers would have to make different recalibrations, depending on where their gauges were to be sold. If someone purchased a recalibrated gauge in one area and used it in another, its readings would be incorrect. Such a situation would be a local gauge transformation that does not correspond to an invariance. If the International Society of Gauge Manufacturers (I don't think such a society exists) wanted to prevent such erroneous measurements, it would have to attach a corrective device to each gauge that sampled the local air pressure and made the correct recalibration for each location. Such a positionally dependent corrective device, whose sole reason for existence is to insure the local gauge invariance, is called a "gauge field."

In modern field theory, local gauge transformations are made to be symmetries by adding terms to the equations that correct for the changes imposed by the transformations. These added terms are also called "gauge fields," and they turn out to have extraordinary physical significance. In fact, they produce the forces among particles and are themselves a reason for the existence of certain classes of subatomic particles.

Of course, I have just made things very simple. Subatomic physics is not quite the same as the theory of tire pressure. For one thing, the symmetries that are purported to exist in the subatomic realm are not found in the same manner as simply rotating a physical object in three-dimensional space and checking to see if it looks the same after it is rotated. There is a rotation, but the "object" that is rotated is a certain set of particles, and the "rotation" turns the particles into each other.

Five

Frames of Reference

The early Greek philosophers, including Plato and Aristotle, believed that Earth was special because we were put here. They also believed heaven was special because it was the home of the gods. But because of the significant distinction between men and gods, heaven was quite different from Earth. Consequently, Aristotelian science consisted of two parts, a physics to explain behavior on Earth and an astronomy to explain behavior in the heavens.

In Aristotle's astronomy, the planets moved in circular orbits because circles were "perfect" shapes (having neither beginning nor end) and the gods would naturally demand perfection. Here on Earth, however, things moved in straight lines, because the linear form (having a beginning and an end) was imperfect, as befit our own imperfections. It was clear to Aristotle that there was no symmetry with regard to one's place in the universe. If, while your eyes were closed, you were transported from Earth to the heavens, you would clearly know a change had occurred.

Without going too deeply into the history of science in the fifteenth and sixteenth centuries, let me simply point out that four of its leading practitioners, Nicholas Copernicus, Tycho Brahe, Johannes Kepler, and Galileo Galilei, created a new cosmic order by noting that Earth was not as special as had been thought (according to Copernicus it was not the center of the universe, and according to Galileo it was not the only planet with moons) and the sun, presumably a perfect heavenly body, was not as perfect as had been thought (it had sunspots). Shifting Earth from its central position in the cosmos to a peripheral position imposed a new symmetry on nature: the position of Earth in the cosmos was unimportant. Nature was indifferent to your place in the cosmos. Of course, the distinction between Earth and the cosmos still held; it was merely the position of Earth within the cosmos that had become irrelevant.

Although the Copernican revolution was decisive in its creation of the solar system, the subsequent Newtonian revolution, about a century later, demonstrated that heavenly phenomena could be explained with the same laws and theories as earthly phenomena. Thus, along with the symmetry that Earth's position in space was immaterial, Newton added an additional symmetry that Earth and the cosmos were also indistinguishable. From the standpoint of the laws of physics, not only didn't it matter where in the cosmos Earth happened to be; it also didn't matter if you were on Earth or elsewhere in the cosmos. One of Newton's greatest contributions was his creation of the universal law of gravity, a simple description of gravitational attraction that allowed a single explanation of both the closed orbits of the planets in heaven and the straight-line motion of the falling apple on Earth. Newton in effect expressed the new symmetry of space in mathematical terms, devising formulae that were completely independent of place.

Inertial Reference Frames

To students of physics, Newton is associated with two major contributions, the law of universal gravitation and the three laws of motion. Of the three laws, it is the second that is usually remembered, because it is the one on which all the problems at the end of the chapter in textbooks are based: $F = ma$. Force equals the product of mass and acceleration. It was this law, in fact, that enabled Newton to deduce the law of universal gravitation, so $F = ma$ is doubly important. I do not intend to dwell on this law, however. Instead, I will discuss the first law, which Newton stated in the following words:

> Every body continues in its state of rest, or of uniform motion in a straight line, unless it is compelled to change that state by forces impressed upon it.

This is called the law of inertia, and to many it seems to be no more than a special case of the second law, which would arise when the F in $F = ma$ is zero. But in fact the first law is there for a purpose. Newton did not propose it gratuitously. It is Newton's attempt to replace Earth as a special frame of reference with a new special frame of reference, in which objects not acted upon by forces move in straight lines. Earth now loses whatever claim it had to being special. For Newtonian physics, it is not Earth that is important, it is this new frame, called an inertial reference frame. This is by no means a trivial point. In Aristotle's theory of nature, there were only two frames of reference: Earth and the heavens. Humans had no choice which to use. The gods were granted the perfect frame, and as a result they alone had insight into the truth. Humans had to view nature from below, never being able to attain the insight of the gods, but striving nevertheless to see as clearly as possible. With the unification of Earth and heaven by Copernicus and Newton, all inferior and superior frames are abandoned.

It is as impossible to remove reference frames from a physical description of natural phenomena as it is to write a novel without a language. The question of whether there are still preferred reference frames therefore is a critical question. One can just as well ask, Are there preferred languages for expressing certain ideas? Is there a preferred medium for artistic expression? Is there a preferred method for filming a certain scene? For physics, at least, Newton answered that question in the affirmative with his first law. The preferred reference frames were those in which an object not acted on by any forces will move in a straight line with constant velocity.

The rotational motion of Earth gives rise to two important noninertial effects that make themselves evident as additional "accelerations" when we observe the motion of objects, centrifugal and Coriolis accelerations. It is a tribute to Newton's perceptiveness that, even though he had always been an earthbound observer himself, he could nevertheless conceive of the existence of inertial frames of reference. Newton's final assertion is that all inertial reference frames (remember, he had never experienced even a single one) are equivalent. He proved this mathematically by showing that if his second law of motion was correct in one such frame, it remained true in any other frame that moved with a constant velocity relative to that frame.

By the beginning of the eighteenth century we have, therefore, the general belief that the universe is symmetric with respect to position, that inertial reference frames are the preferred frames, and that these frames are symmetric with respect to interchange (there is no detectable difference among inertial reference frames). In short, if two physicists, both using inertial reference frames, were to perform all the pertinent experiments of physics and induce from them the appropriate laws of nature, they would discover that they had derived identical laws. Nature is indifferent to one's choice of inertial frame, although it is not indifferent to a choice between an inertial frame and a noninertial frame.

This belief persisted until the middle of the nineteenth century, when new experimental data began to shed doubt on the equivalence of all inertial frames.

The Ether

From the 1850s onward, the emphasis in physics shifted from the investigation of mechanical phenomena through Newtonian physics to the study of electrical and magnetic phenomena, which were possibly not described at all by Newton's system of three laws. One of the phenomena that came under close scrutiny was the propagation of light, although it was not at first clear that light was an electrical or a magnetic phenomenon. In Newton's time light was a real mystery, and there was even disagreement as to whether light was some kind of wave motion, like sound or waves in water, or particle motion, like a succession of little BBs. By the nineteenth century, however, experiments had clearly demonstrated that light was a wave.

Now a wave is simply a disturbance that moves through a medium at a characteristic speed. Sound, which is the classic example of wave motion, is a disturbance in air (or other solids and gases) that moves with a speed determined by the density and temperature of air (about 1,100 feet per second under normal conditions). The behavior of light was thought of in a similar fashion. The accepted theory was that there had to be a medium for light, like the air was for sound, whose movement became the light itself and whose properties (like density) determined the speed of light (186,000 miles per second or 300,000,000 meters per second). The problem was that this medium would have to have some rather amazing properties. For one thing, it would have to completely fill the universe, since light obviously propagated between the stars (where there was certainly no air to propagate sound). In addition, this medium would have to be capable of flowing through solid objects like glass, through which light also penetrated. With all of these features, the medium would have to be completely imperceptible by us—have no smell, taste, or feel—do nothing but carry light. This supposed medium was called "ether."

If the ether really existed—and this was never seriously in question—a severe blow would be dealt to the equivalence of all inertial reference frames, because it could be easily shown mathematically that any inertial reference frame that happened to be at rest in the ether would have a preferential view of light propagation that no other inertial frame would have. This special status of certain reference frames is, of course, well known in the case of sound propagation, where "still" air is a preferential medium as compared to moving air. Consequently, an observer at rest in the air is a preferential observer for sound as compared to an observer who is moving through the air. Sound of a certain frequency traveling through still air will be perceived as having different frequencies and speeds by observers who are themselves moving through the air. This well-known phenomenon is called the Doppler effect. By the same reasoning, if a beam of light is traveling through the stationary ether with a certain frequency, it should be observed as having different speeds and frequencies by inertial observers that are moving with respect to the ether. Thus, of the infinity of Newton's equivalent inertial frames, all of which were supposedly special, only one, the "ether-rest frame" would remain special. The other inertial frames might still be equivalent with respect to mechanical phenomena, but they were certainly rendered distinguishable by the behavior of light.

There were other reasons, too, for judging the ether-rest frame to be special. By 1873, the English physicist James Clerk Maxwell had successfully incorporated all of the experimental results of electricity and magnetism into a single set of equations that have come to be called Maxwell's equations. Not only did these equations serve as a powerful unifying theory, incorporating and consolidating more than a hundred years of diverse experimental evidence gathered by such eminent scientists as Coulomb, Ampère, and Faraday, but they also predicted new phenomena. In particular, although Maxwell did not incorporate the phenomenon of light into his equations, it came out of them. The equations, when solved, provided a wavelike phenomenon that moved with precisely the speed of light. This miraculous result cannot be overemphasized. Physicists now had complete justification for considering light to be electromagnetic in origin. Without Maxwell's equations, light could equally well have been an entirely new phenomenon, subject to laws of its own.

Maxwell's equations took on a characteristic mathematical form when expressed in a frame of reference in which the batteries, wires, electrical charges, and magnets that produced the electrical and magnetic effects were at rest. This frame was therefore presumed to be the ether-rest frame, even though no explicit reference to the ether appeared in the equations (with the exception of certain constants that had been measured experimentally and were believed to be a result of the presence of the ether). When Maxwell's equations were recast in a form appropriate to a reference system moving "through the ether" (this is done by means of what mathematicians call a transformation of coordinates in the equations), they assumed a far more complicated form, acquiring many extra terms that depended on the speed of the moving system. This change in the form of Maxwell's equations was taken to be evidence of the ether's presence, and the extra terms were considered to be effects of the ether that could be measured with sufficiently careful experiments. It is important to point out that when Newton's second law, $F = ma$, is expressed in terms of the coordinates of a moving reference frame, it still comes out as $F = ma$. No extra terms appear. This is indicative of the equivalence of all inertial frames for phenomena that are described by $F = ma$.

Since expressing Maxwell's equations in a moving frame of reference generates new terms that are in principle measurable, it simply remained for some clever scientist to figure out the proper experiment with which to make the measurement. Many scientists accepted the challenge. Most of their experiments involved attempts to measure changes in the behavior of light as it traveled through moving media, such as flowing water. Ultimately it became clear that none of these experiments in fact detected any differences in the behavior of light. Various explanations were found to explain the lack of results, but the conclusion seemed to be that no experiment was able to detect the effects of motion through the ether.

In 1881, the American physicist Albert A. Michelson believed he had conceived exactly the right experiment: he would measure the speed of Earth itself through the ether by using the effect of that motion on the speed of light. The basis of this experiment was Michelson's assumption that Earth, given its complicated orbital motion with a speed of about 30 kilometers per second, is clearly not the ether-rest frame. Therefore, Earth's speed through the ether, whatever that happened to be, would have to produce measurable effects on the speed of light.

Much to Michelson's dismay, when he performed his carefully conceived experiment, his results implied that Earth's velocity through the ether was zero. The extra terms in Maxwell's equations that were supposed to be there, simply weren't. At least he couldn't measure them. Six years later, in 1887, Michelson teamed up with another physicist, Edward Morley, and redid the experiment using much more precise and sensitive measuring devices. Still he had no luck, getting the same answer: zero.

The device used by Michelson and Morley to detect Earth's motion through the ether was extremely sophisticated and sensitive, so much so that Michelson eventually won a Nobel prize for its invention, becoming the first American to win that prestigious award. Notice that he didn't win the prize for the experiment; he won it, in essence, for the speedometer, which was called an "interferometer" (now called a Michelson interferometer).

The basic idea of the experiment was as follows. On the assumption that Earth is moving in some direction through the ether, a beam of light shining from Earth in the direction of its motion will behave differently (from the standpoint of an earthbound observer) from an identical beam shining at a right angle to the first beam. To the observer, the beam moving into the ether is somewhat impeded in its movement away from Earth (like shouting into the wind), whereas the beam moving at a right angle is not affected in the same way. To an observer at rest in the ether, both beams are moving away from Earth at the same speed, but Earth is continually moving in the direction of the forward-moving beam, in effect chasing after it. The problem is, how does an observer on Earth go about measuring the progress of the two beams of light? Michelson and Morley's ingenious method was to use the properties of light itself to make the measurement.

Instead of simply sending two beams at a right angle to each other out into the void, never to return, Michelson's interferometer was a device in which mirrors reflected each beam of light back to its point of origin. The beams thus made a round trip. However, the device didn't use two separate beams, like those that would be produced by two flashlights; rather, it used a single beam that was split in half, each part moving off at a right angle to the other. This insured that the two beams, in terms of their wavelike properties, were perfectly in step with each other when they were sent off. The interested reader is urged to visit an institution that has a working model of the interferometer, as it is quite interesting to watch in operation.

Michelson and Morley intended to measure the time it took for each beam to complete its round trip and to demonstrate that the two times were different. It was not particularly difficult to calculate the amount of time by which these trips should differ, which was miniscule—so miniscule, that no clock then existing could measure it. However, the light wave itself acts as a clock. Its vibrational frequency corresponds to a ticking rate. Since each beam returns to its point of origin, the beams could be superimposed on each other. If one beam took more time for its trip, the peaks and valleys of the two waves, which were exactly in step when they began their trip, would be out of step on their return. The amount by which they would be out of step would be the difference in the number of ticks made by each wave in its round trip. Laying one wave over the other in just the right way would produce a pattern that could be measured, from which the time difference could be calculated.

When the experiment was performed, however, Michelson and Morley found no measurable difference. The two waves were as perfectly matched when they returned as when they were sent out. It would be an understatement to say that this result caused consternation within the physics community. Not surprisingly, many explanations were offered for the failure of the experiment. One explanation suggested that Earth, because of the roughness of its surface, actually carried a stagnant layer of ether along with it, so that the experiment was actually measuring the lack of motion of this layer rather than Earth's motion through the ether in space. To explore this possibility, Michelson redid the experiment at the top of a mountain, where the stagnation effect would be significantly reduced. The result was the same: no detectable motion.

Perhaps the most ingenious explanation was offered independently and simultaneously in 1892 by H. A. Lorentz and G. F. Fitzgerald. They noted that the time difference Michelson and Morley were attempting to measure depended on the assumption that both arms of their interferometer, the flight paths of the light beams, were the same length. If unbeknownst to them the arms were of different lengths, the time difference would not be what they assumed. Suppose, for example, that the actual length of the arm that points into the ether were less than they assumed by exactly an amount that would cancel the effect of the ether. This would be like trying to measure the speeds of two runners by having them race on different tracks, presumed to be identical in length. If the two runners do a lap and return at the same time, it could be said that their speeds were equal. But if one track was found to be shorter than the other, the runner on that track would have actually been slower.

Now the problem with the short-track analogy is that one could correct things by having the runners switch tracks. Indeed, in the Michelson-Morley experiment one merely had to reverse the positions of the arms to avoid this problem. In fact, this is what actually happened in the experiment: the arms were rotated as a matter of course to correct for possible errors. But the Lorentz, Fitzgerald hypothesis (called the Lorentz-Fitzgerald contraction) was cleverer than that. It theorized that the arm pointing into the ether, whichever it is, is made shorter by its very motion into the ether. This was as though the track on which the slower runner was racing would always be shorter, regardless of which track it was!

The beauty of this hypothesis is that it is irrefutable. The arm shortening could neither be proved nor disproved by any independent measurement. If you tried to verify lengths with a ruler, the ruler was also shortened. It could even be shown that you couldn't see the length of contraction with the naked eye, because the eye also changed its dimensions in such a way as to make the contraction of the arm undetectable. The only way you knew the contraction had occurred was by the fact that you could not detect the time difference. Of course you knew there was a time difference, because you knew there was an ether.

What causes this miraculous contraction? It is caused by the ether itself. The ether, of course, mediates all electrical and magnetic forces. It is electrical forces that are responsible for the distances between atoms in all materials. Thus it is not unreasonable to suppose that when atoms move through the ether, the forces between them change in a way that allows the atoms to move closer together and to therefore shorten the very object that the atoms construct.

Let me conclude this bit of history by saying that there is nothing wrong with this entire point of view. It is possible to hypothesize that the universe is indeed filled with an ether and that this ether affects interatomic forces in such a way that objects are shortened along the direction of their motion through it. Furthermore, as a result of this shortening it becomes impossible to measure the speed of Earth through the ether, even though all reasonable people know that it is indeed moving. As "reasonable" as all this is, however, it is a hypothesis that has been totally rejected and replaced by another. This new hypothesis, called special relativity, requires assumptions that are even stranger than the Lorentz-Fitzgerald contraction. Its beauty, however, is that it can be cast in terms of a symmetry and it leads to an incredible simplicity of our physical conceptions. We are not waging a battle between truth and falsehood; rather, we are confronting the complexity of one hypothesis with the simplicity and elegance of another.

Einstein's Theory of Special Relativity

The entire edifice of ether theory was dismantled in 1905 by Albert Einstein's theory of special relativity. It would be tempting to speculate that Einstein studied the Michelson-Morley experiment and the Lorentz-Fitzgerald contraction hypothesis and tried to explain the negative result of the former without resorting to the latter. Unfortunately, and unbelievably, it is not clear that Einstein had even heard of the Michelson-Morley experiment and the Lorentz-Fitzgerald contraction hypothesis when he created relativity. He was virtually out of the mainstream of physics; he couldn't even get a job teaching physics after graduation, and he was working as an examiner in the patent office in Bern, Switzerland. It is doubtful that Einstein had access to the physics journals that contained the Michelson-Morley results. In any event, as we shall see, their results and indeed their entire experiment were virtually irrelevant to Einstein.

At this time in his life, Einstein had two overriding philosophical passions. He had been very strongly influenced by the "positivist" philosophy of the nineteenth-century Austrian scientist Ernst Mach, who is today best remembered for the "Mach number," designating multiples of the speed of sound. Mach was an outstanding philosopher, and his version of positivism declared, among other things, that only what you measure is real and that all theories in science (and, indeed, in other disciplines as well) should concern themselves only with measurable quantities. To Mach, as to any positivist that believed as he did, the notion of an ether was absolutely laughable. If you couldn't see it, smell it, taste it, or feel it, it had no place in a theory, let alone in a central position. It is quite certain that Einstein shared this low opinion of the ether and would feel no great commitment to its preservation. In contrast to a disbelief in an ether, Einstein would have had a strong general belief in the value of measurements and the results of experiments, particularly those that had attempted and failed to measure the speed of light relative to Earth, which the Michelson-Morley experiment was trying to do.

No self-respecting positivist would throw out good experimental results because they betrayed some "theory." Consequently, Einstein accepted what, to him, was the simplest and most obvious hypothesis, namely that differences in the speed of light due to the motion of an observer were, in fact, unmeasurable. Let me add that Einstein would have been thinking in terms of all the other experiments done prior to those of Michelson and Morley, which had also found no effects of motion through the ether. Einstein clearly knew of those results, even if his knowledge of Michelson and Morley's experiment is in doubt. Einstein's bold, even audacious, stroke of genius was to take as a fundamental hypothesis an observation that to everyone else required a complicated explanation. Einstein began with the assumption that no differences in the speed of light can be measured.

Einstein's other philosophical passion was his belief in symmetry, the symmetry among inertial reference frames, to be exact. Quite simply, Einstein believed that all inertial frames should be equivalent and that therefore any physical theory put forth as being correct had to pass the symmetry test: the equations of this theory must have the same form in all inertial reference frames.

As strongly as he believed in the importance of measurable quantities and the symmetry of inertial frames, Einstein also believed in the correctness of Maxwell's equations. Now this seems totally contradictory, because if one thing was clear, it was that Maxwell's equations didn't look the same in all inertial reference frames. So you couldn't believe in Maxwell *and* in the symmetry of inertial frames. Or could you?

Einstein was already convinced by his experience as a physicist—that is, from his knowledge of the results of certain basic experiments—that Maxwell's equations had to be the same in all inertial frames. What made it seem otherwise was the purely mathematical demonstration that when those equations were expressed in terms of the coordinates of a moving inertial frame, they changed their form and acquired additional terms. You might recall the additional terms that appeared when the equation of a circle was shifted by five units. Thus, in Einstein's mind there were competitive claims between his experience, which said equivalence, and mathematics, which said inequivalence. Given Einstein's positivist sympathies, there was no question of which claim would win.

I should say a few words about why Einstein was so firmly convinced that Maxwell's equations had to be the same in all inertial frames. He begins his revolutionary 1905 paper "Zur Elektrodynamik bewegter Körper" ("On the Electrodynamics of Moving Bodies") with precisely that point. I will paraphrase his explanation. Although experience shows that moving a wire in the presence of a stationary magnet or moving a magnet in the presence of a stationary wire produces exactly the same effects on the electrons in the wire (i.e., a force), physicists tend to explain the two cases differently when using Maxwell's equations. When the magnet is stationary, one can use Maxwell's equations in their ether-rest-frame form—that is, in their simplest form. When the magnet is moving, however, one has to use the form of Maxwell's equations appropriate to a moving inertial frame, because only in that frame is the motion of the magnet correctly accounted for. Now these two versions of Maxwell's equations are quite different, yet the actual phenomenon is the same in both cases and should be entitled to a single description.

To Einstein, the only thing that is physically measurable, certainly the only thing that would be felt by an electron, is the relative motion between the magnet and the wire. Therefore, (to the electron) the two phenomena are really a single phenomenon, even though superficially (to the physicist) they seem different. Einstein takes the electron's point of view as being more basic than the physicist's. The single phenomenon is simply a magnet and wire in relative motion.

The physicist who believes in the ether's existence sees two quite different phenomena. In one, a magnet moves through the ether, and in the other, a magnet is at rest in the ether. The physicist reasons that two mathematical descriptions should be necessary. To Einstein, this difference is important only if there is a stationary ether, which his positivistic philosophy does not give credence to. What Einstein sees as important, on the other hand, is the relative motion, which is observable and measurable whether or not an ether exists. The relative motion being the same in both cases, however, tells Einstein that there is only one phenomenon, so you'd better get the same description however you choose to use Maxwell's equations. Therefore, if Maxwell's equations turn out to look "different" in a moving frame, you must be making a mistake in the way you change the equations to fit the moving frame. It is not the equations that are wrong, nor is it the concept of the symmetry; it is simply the way you make the mathematical transition from one frame to another.

Einstein had found a fundamental error in the way things were being done. The mathematical transformation between one inertial frame and another one, the so-called symmetry transformation, was incorrect. The question was, How to find the correct transformation? This also raised a rather disquieting secondary point. The transformation that Einstein felt was in error was the same transformation used successfully by Newton to prove that his second law, $F = ma$, was the same in all inertial frames. This transformation, which had come to be called a "Galilean transformation," had had a long history of successes and, not surprisingly, had been applied unhesitatingly to Maxwell's equations. The Galilean transformation was expressed in a series of four equations:

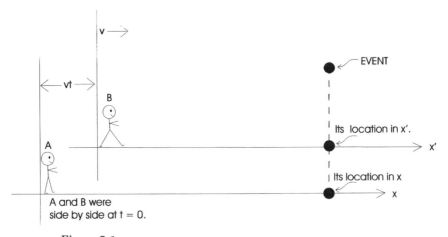

_____ *Figure 5.1* _____
Galilean Transformation

(1)
$$x' = x - vt;$$
$$y' = y;$$
$$z' = z;$$
$$t' = t.$$

These four equations express the relationship among the space and time coordinates of the same event, as determined by two observers moving relative to each other with velocity v along the x direction. Assume one observer, called (B), is moving in the positive x direction with velocity v relative to first observer, called (A). If observer (B) looks at an event that observer (A) claims is located at x, y, z, t, then observer (B) will claim that the same event is at x', y', z', t'. The primed variables used by (B) are related to the unprimed variables used by (A) by the four equations (see fig. 5.1).

If this transformation was wrong, then a new and "correct" transformation would presumably destroy the symmetry of Newton's law. Could it be that the symmetry among inertial frames was true for electromagnetism but not for mechanics? We shall see how Einstein handled that conundrum.

The simplest way to express the symmetry among inertial frames is to say that any physically possible electromagnetic phenomenon must appear to be so to all inertial observers. Note that a symmetry does not require that each individual phenomenon look exactly the same to every inertial observer. That would be more than just a symmetry; it would be an "identicality," to coin a word. Symmetries of nature do not require individual identity on a phenomenon-by-phenomenon basis; rather, they require only that something that appears to be a reasonable phenomenon to one observer remain reasonable to another. In this way, no inertial observers could use some particularly weird observations to delineate their states of motion.

Einstein set about to determine the transformation that would preserve the form of Maxwell's equations in all inertial reference frames. He did it in a fascinating way, not by studying the equations (which are quite complicated), but by imagining a simple hypothetical experiment, whose outcome he believed he knew precisely. The experiment is a simple one. A single pulse of light is produced by popping a flashbulb somewhere in outer space. The pulse is noted by two inertial observers, moving relative to each other, one of whom is at rest relative to the bulb. In the inertial frame in which the bulb is at rest, the pulse of light is an expanding sphere whose radius grows at the speed of light. The mathematical expression for such a growing sphere is

(2) $$x^2 + y^2 + z^2 = c^2 t^2,$$

where c is the speed of light. The coordinates x, y, and z refer to any position on the expanding spherical pulse.

The second inertial observer, moving relative to the first one but passing by just as the bulb goes off, looks at the same expanding sphere of light and describes it in the same mathematical form but a different set of coordinates, which we'll label (x', y', z', t'). This means that the equation of the sphere to the moving observer is

(3) $$x'^2 + y'^2 + z'^2 = c^2 t'^2.$$

It is crucial that the reader understand, just as Einstein did, why the moving observer must see the same spherical form for the expanding pulse of light: the moving observer measures the same speed of light in all directions that the stationary observer does. This is one experimental result that Einstein was certain of.

On the face of it, the result of this experiment seems utterly ridiculous. It says that each observer sees a sphere of light centered on that observer's own coordinate origin. Yet we know that the flashbulb that produced the light is at rest in the first inertial frame, and surely it is only in that frame that the sphere is centered on the origin. The second observer continues to move away from the first. Clearly, the second observer has to see a sphere of light centered on the flashbulb, which remains with the first observer. Throw out your intuition! This is how it must be, said Einstein, if the results of experiments measuring the speed of light are correct and if the symmetry is correct. In fact, this experiment is not different from Michelson and Morley's, if we think of the flashbulb as being mounted on their interferometer.

Now we must ask what is the correct transformation of coordinates that will allow each observer to describe the same sphere (forgetting for a moment that we are disinclined to believe the entire result). I will simply state the answer:

(4)
$$x' = (x - vt)/(1 - v^2/c^2)^{1/2};$$
$$y' = y;$$
$$z' = z;$$
$$t' = (t - vx/c^2)/(1 - v^2/c^2)^{1/2}.$$

This is a very strange set of formulas, quite different from and much more complicated than the Galilean transformation of equation (1). It has one crucial virtue, which, of course, it was designed to have. It allows both observers to describe the sphere of light in exactly the same way. You can prove this quite easily with a bit of algebra. Just substitute the formulas for x', y', z', and t' given above into the sphere equation (3). You'll obtain the identical sphere equation, but in terms of x, y, z, and t, exactly as given in (2). In other words, the invariance is assured by the transformation in (4).

The price you must pay to retain the observer-centered sphere in both frames is an interrelation between spatial coordinates and time that seems to totally contradict intuition. For one thing, the relationship between t and t' indicates that the clocks of the two observers will begin to disagree on the times at which things happen. They agree only at the moment they pass each other and are at the same position: $x = x' = 0$. At that precise position the observers can set their clocks to the same setting, which we will call $t = t' = 0$.

Einstein saw that he was going to have trouble selling this new transformation to the scientific community. In fact, he could see even more nonintuitive results arising from his transformation than I have just mentioned. He thus decided that, in order to give the transformation a semblance of rationality, he would have to attack and demolish all intuition regarding space and time and rebuild it from scratch. Such a rebuilding would not only make the transformation believable; it would also make reasonable the fact that both observers see a sphere of light centered upon themselves. Einstein's 1905 paper, in which this transformation appears, contains precisely this attack.

The core idea of Einstein's rebuilding of our intuition is that we must stop thinking about "space and time," which is something that is "given to us," and must instead think about "measuring positions and times," which is something we do. Only our measurements have real existence. We build up an intuition of something called space and time, which we believe exists beyond these measurements and which would be there even if the measurements were never taken. Einstein's first commandment was to pay attention only to your measurements and worry later about the properties of the more abstract notion of space and time.

A good example of this kind of reasoning arises in the educational sphere when we think about the concept of IQ. What is an IQ? It is really only a number we get when we grade a particular examination. Yet it has taken on a conceptual life of its own, becoming confused with "intelligence," a much more nebulous and abstract concept. Most people think that a way to "improve" IQs is to improve education. But another way would be to change the test! I don't intend to get involved in an argument over IQ versus intelligence, but the fact that such arguments are common is an indication of the degree to which we confuse measurements with more abstract notions that we build up around them.

Six

The Einsteinian Attack on Space and Time

Although Einstein's coordinate transformation preserved the symmetry among inertial reference frames, it was at the price of creating unusual relationships among the positions and times of identical events as recorded by the clocks and rulers of different inertial observers. To make these relationships as plausible as possible, Einstein had to revise traditional ways of thinking about the nature of length and time duration. True to his positivist leanings, he wanted to restore the importance of the actual measurement process to the way in which length and time were defined. To Einstein, "length" was not an innate characteristic of an object; it was the result of a measurement obtained by a very specific process involving an observer, a ruler, some clocks, and the object. The same was true for time duration.

Einstein began his explanatory task by asking what is probably the most fundamental question of all: How do you measure the length of an object? The usual answer goes something like this. Place a ruler alongside the object whose length is to be measured, and note the numbers on the ruler that coincide with the ends of the object. Subtract the smaller number from the larger. The result is the length of the object.

Now this technique is perfectly appropriate when the object being measured is at rest relative to the ruler, but it will not work if the object is moving. If the ruler is placed so that first it is alongside the front of the moving object and that number is noted, and then the number alongside the rear of the object is noted, the subtraction of the two numbers will produce a length that is shorter than the value obtained for the nonmoving object. The reason, of course, is obvious. During the inevitable delay between noting the number alongside front and rear, the object has moved. It is precisely by the amount of this motion that the measurement will be in error.

The resolution of this problem is to place the ruler alongside the moving object so that both ends of the object are coincident with the numbers on the ruler at the same time. Although this seems obvious, it is actually a radical proposal. It implies that the measurement of a length fundamentally involves one's judgment of a temporal quantity, the simultaneity of ruler placement. In and of itself, this hardly seems to be earthshaking. It is all very well to make prospective length-measurers respect the importance of time in the measurement process. But here is where Einstein makes his most revolutionary proposal. He will demonstrate that the very concept of simultaneity, the observation of events occurring at the same time, is not absolute. What is simultaneous for one inertial observer is not simultaneous for another moving relative to the first. The concept of simultaneity, which is fundamental to the definition of length, must itself be completely rethought.

Simultaneity

Why can't two inertial observers in uniform motion relative to each other agree that two things happen at the same time? Surely "happening at the same time" should not depend on one's point of view. But it does. And the reason it does is that the speed of light is the same to all inertial observers. And why is the speed of light the same to all inertial observers? That's a very good question, and one good answer is that nature just behaves that way! Einstein might say that if the speed of light weren't the same for all inertial observers, it would provide a mechanism to determine your absolute speed and to determine which observer is "really" moving. This, of course, would be unacceptable to a believer in symmetry. I could be somewhat more technical and say that the speed of light is the same to all inertial observers because Maxwell's equations are the same to all inertial observers, and it is those equations that describe the motion of light and correctly predict its speed. Of course, one could then innocently ask, "Why are Maxwell's equations the same for all inertial observers?" To which I would once again respond, "Nature just behaves that way!" Einstein didn't dwell on these "why" questions; he simply accepted the invariance of the speed of light as a fundamental fact of nature and went on to demonstrate its consequences.

Let me begin this explanation of how and why the properties of light impose themselves on simultaneity by using a simple analogy. Imagine you are observing a distant thunderstorm. You see a stroke of lightning and, shortly thereafter, hear the ominous rumble of thunder. Someone standing nearby asks you, "Do lightning and thunder occur simultaneously?" "Of course they do," you answer. "Then why is there always a time delay between them?" comes the next question. "Because," you respond wisely, "the speed of light is 186,000 miles per second, whereas the speed of sound is a mere 1,100 feet per second—about a fifth of a mile per second. Thus the light reaches you virtually instantaneously, while the sound moves at a much more leisurely pace." This, of course, is the correct explanation. But suppose it turned out that the speed of sound and the speed of light were actually equal. Suppose you found out that all these years you had been wrong in assuming they were different. After all, you had never actually measured the speeds of light and sound; you had always relied on the word of others (your physics teacher, for example), and we know how notoriously unreliable the word of others can be. If it turned out that the speeds were identical, then seeing the lightning first and hearing the thunder later would mean that in fact the lightning occurred before the thunder. In short, your judgment of when things happen depends on the speeds of the signals that carry the information of their occurrence to you. There are very few things that happen so close to us that the speed of the signal is of no importance.

Just as finding out that the speeds of light and sound were the same would cause you to modify your judgment of when things happen, so too would finding out that the speed of light never changed, regardless of how fast you were moving. Einstein himself used the following argument to make this plausible. Imagine you are on a train, standing precisely at its center. The train is moving with a speed V (see fig. 6.1) relative to the ground. Your friend happens to be on the ground, standing by the track as your train is passing. At the instant that you are alongside him and he sees your smiling face in the window, two bolts of lightning strike at the front and rear of the train and are seen at precisely the same time by your friend. The bolts leave noticeable marks on both the train and the track indicating where they struck. You and your friend decide to determine independently whether or not the two bolts struck simultaneously.

This is reasonably simple for your friend to do. He carefully measures the distance from each mark left on the track to the place where he stood and verifies that the distances are equal. He also knows that the speed with which the light from each bolt reached his eyes was the same. (He could measure this if he cared to.) By combining the factors 1) same time of observation, 2) equal distance traveled by the light, and 3) same measured speed of light in both directions, your friend at the side of the track can easily conclude that the bolts actually struck simultaneously.

The situation for you on the train is a bit different. In fact, until we examine the accompanying figure, we do not even know if you see the two bolts simul-

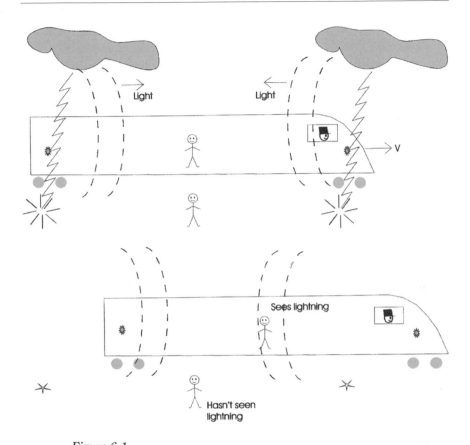

_____ *Figure 6.1* _____
Simultaneity on a Train

taneously, let alone whether you actually conclude that they struck simultaneously. By examining that figure (which is a luxury we have, because we can observe everything that is happening), we can conclude that you do not see the bolts simultaneously, because you are moving toward the light produced by the forward bolt and away from that produced by the rear bolt. In fact, you see the forward bolt a short time before your friend does, and you see the rear bolt a short time after your friend does. This is an objective truth that does not alter the experimental fact that you will measure the speed of the light that approaches you from the front and from the rear.

Now you have to decide when the bolts actually struck the train. First you measure the distance from the place where each bolt struck the train to your own position on the train. You conclude that you were exactly halfway between the two bolts when they struck. Finally, to be thorough about this, you verify the speed with which the light from the bolts reached you. If you believe Einstein's assertion and Michelson and Morley's result and if you actually did the experimental observation yourself on the train, you would find that each beam moved toward you with the correct speed of light. Therefore, you must conclude that the two bolts struck the train at different times. The conclusion follows inescapably from the following chain of reasoning. You saw the bolts at different times. You were exactly halfway between them when they struck, and their light traveled toward you at the same speed. Therefore, they must have struck the train at different times.

One point about this experiment sometimes causes confusion. Why would you, on the moving train, persist in believing that the bolts struck at the positions (also moving) that are marked on the train? After all, isn't it the fixed marks on the ground that really determine the places where the lightning struck? The answer is that for you on the train, the train is your world. You must make all your judgments of events based on points of reference in that world. This is what it means to be an "inertial observer." I could have made a more convincing argument by requiring that there be no windows on the train and that it be moving so smoothly that you do not even perceive your motion. The lightning bolts then intrude on your world by suddenly causing two bright flashes and the appearance of noticeable marks. You do not think to yourself that the bolts "really struck outside" but happened to be seen by you inside; you think only that the two bolts struck within your own world. All observers thinks their world the center of all things and make judgments based on that belief.

The disparity between the conclusions of the two observers is striking. These results would not have been even imagined before the Michelson-Morley experiment or Einstein's paper, simply because nobody would have had any reason to believe anything other than that the speed of light would be increased or decreased by the amount of one's own speed. The two observers, knowing that light moves through the ether, would have each believed that they would measure different speeds of light, had they the necessary apparatus. The stationary observer, supposedly at rest relative to the ether (actually on the moving planet Earth, but let's imagine it stationary in the ether), would firmly expect to measure 186,000 miles per second in both directions. The moving observer would just as firmly expect to measure two different speeds of light, $186,000 + V$ in the direction of motion and $186,000 - V$ in the opposite direction. If this were actually the case, the moving observer could legitimately conclude that the bolts struck simultaneously. The fact the forward lightning bolt is seen first and the rear bolt a bit later would simply be a result of the different speeds of the light relative to the moving observer. In short, the situation is remarkably like the lightning and thunder analogy discussed earlier.

The "raw" observation by the moving observer requires only the observer's eyes, which suggest that the bolts strike at different times. The conclusion drawn from that observation requires the additional knowledge that the speed of light is 186,000 miles per second (or 3×10^8 meters per second) regardless of the observer's motion and is the same in both directions. That requires a firm belief in Einstein's hypothesis. Clearly, the relativity of simultaneity is not a discovery, because nothing new has been "discovered"; it is a creation of Einstein or of Michelson and Morley. You might say, "Why does it require a belief in Einstein's theory? The observer can certainly measure the speed of light and see that it is the same in both directions." True enough, but if the observer believed in ether theory, the mere measurement of the same speed would only mean that the measuring devices were being tricked. The point is that the times of observation of the lightning bolts are the same before and after Einstein; it is the conclusion as to when they actually strike that changes radically. The observation involves your senses; the conclusion requires your intellect.

Length Contraction

Having thus established the relativity of simultaneity, we can show that length too becomes relative. When two observers moving relative to each other measure the length of the same object, they must each simultaneously observe both ends of the object alongside the markings on their rulers. However, a simultaneous ruler placement to one observer will not be simultaneous to the other observer, so they will disagree on the result and validity of their measurements. If one of the observers is at rest relative to the object whose length is being measured, while the other observer moves relative to it with a velocity v, their disagreement on the object's length will take the form of a shortening of the length of the object as measured by the moving observer. This shortening, or "length contraction," is by a factor of $(1 - v^2/c^2)^{1/2}$, which is always less than 1, since the v is always less than c, the velocity of light. We shall eventually discuss this length contraction in much greater detail. The observer who happens to be at rest relative to the object will obtain a maximum value for its length, which is called the "proper" length of the object. We might be tempted to call proper length the "real" length of the object, because it corresponds to the length we traditionally measure. However, the use of "real" implies that the length measured by the moving observer is "unreal," illusory, or incorrectly done, and this is just not so. Any length measured by a simultaneous reading of a ruler placed alongside two ends of an object is a legitimate length, in fact is the only length the object can have relative to the observer. That it is less than the length measured by some other observer simply demonstrates that length can no longer be regarded as an intrinsic property of an object, but must also take the motion of the observer into account. Length, in sum, is an interaction between an observer and an object. To discuss an object's length, you must first specify an observer.

The contraction of length deduced by Einstein is commonly called the "Lorentz contraction," in a kind of homage to the Lorentz-Fitzgerald contraction of the ether days. The two contractions are in fact exactly the same magnitude, but have an entirely different origin and interpretation. The Lorentz-Fitzgerald contraction is a physical effect caused by the ether's action on the molecular structure of an object moving through it. It would not have been even thought of if scientists didn't believe that the ether existed. The Einsteinian contraction (Lorentz contraction), on the other hand, follows from the definition of length and the nature of the process by which we measure it. This process requires simultaneity of observations, and therefore involves the invariance of the speed of light. Nowhere is the ether involved! Einstein in fact was noncommittal about the existence of ether, although his positivism would have considered it a meaningless substance, unnecessary for the understanding of physical phenomena. To Einstein, all the effects of relativity result from the symmetry among inertial reference frames, and the existence of any substance that would destroy this symmetry would not be recognized.

Time Dilation

In Einstein's theory, relativity of length is joined by an equally counterintuitive relativity of the duration of time. This additional relative quantity can be deduced mathematically from the Lorentz transformations, and it can be given a more physical justification. In his 1905 paper, Einstein denies that there is any kind of duration that exists in the abstract sense—that is, independently of clocks. He defines duration as a clock measurement, more precisely the difference between two clock readings. It isn't something that preexists and that we measure with a clock; it actually *is* the clock measurement, the distance between ticks. If there are no ticks and no way in principle of producing them, there is no duration.

All real clocks depend on a physical object of a known length for their operation. The time between ticks of a pendulum clock depends on the length of the pendulum. The time between ticks of a quartz clock depends on the vibrational frequency of the crystal, which in turn depends on several of its dimensions (distance between its atoms, the actual length of the crystal, etc.). Even the so-called atomic or molecular clocks, which operate on the basis of the vibrations of molecules or atoms and are the most accurate clocks of all, depend on atomic dimensions for their particular rates of ticking. Given the dependence of length on the motion of the observer, it should not come as a surprise that the rate at which a clock ticks is also observer dependent. In fact, the time between successive ticks of a clock as measured by an observer who is moving relative to that clock will be longer by a factor of $(1 - v^2/c^2)^{-1/2}$ than the time between successive ticks of that same clock when it is at rest relative to the observer. This property is called "time dilation," or "dilatation." The dual properties of length contraction and time dilation are often neatly summed up by the saying Moving rulers are short and moving clocks are slow.

The human heart is also a clock, and it too runs slow when the time between its successive beats is measured by an observer moving relative to it. Thus a person who is moving relative to us will age more slowly than we do, at least in terms of the number of beats his or her heart makes. All other processes associated with age will also slow down, because these too are governed by the ticking of some biological clock. We will discuss this phenomenon a bit later.

Is the Constancy of the Speed of Light a Conspiracy of Clocks and Rulers?

It is tempting to think that the reason all observers who measure the speed of light find it the same is the strange reciprocal behavior of clocks and rulers, a kind of cosmic conspiracy. If you set about to measure the speed of light, you will need to mark off an accurate length and then determine how many ticks your clock makes while a beam of light traverses this length. When you divide the length by the time, you will get 186,000 miles per second. When I move relative to the same beam of light whose speed you are measuring, I have to mark off an equally accurate length in my own reference frame and count the ticks of my own clock. I will also get 186,000 miles per second when I perform the proper division. I know, however, that you can't possibly get the same number I do, because I see you moving into the beam. When you tell me that you also measure 186,000 miles per second, I attribute your "error" to the incorrect length you have marked off with a short ruler and to your use of a slow clock. This may be a tempting way of thinking about things, but it should be avoided. The speed of light is not constant because of a conspiracy of clocks and rulers, but because that is the way the world is. That is the nature of light; nothing more, nothing less. Lengths shrink and times dilate because of the properties of light. Length and time are subordinate to light because they require the properties of light for their definition.

The fact that length contraction and time dilation have a reciprocal relationship does, however, suggest a rather interesting interpretation of spatial and temporal differences in general. It is possible to think about these differences as being very much like what happens to the dimensions of a one-meter rod as it is rotated from a vertical position to a horizontal one. First we might say the rod is one meter "high"; then, we say the rod is one meter "wide." Nothing has really happened to the rod. "High" and "wide" (or "height" and "width") are terms that we find useful for distinguishing between vertical and horizontal dimensions. They are observer dependent and not really attributes of the rod itself. The rod is in fact one meter "long." Its length is its true characteristic; its height or width are simply terms reflecting the perspective of the viewer. Here on Earth, that perspective is useful because there is a clear distinction between horizontal and vertical (thanks to gravity), so everybody knows what you mean when you say high or wide. In outer space, on the other hand, length would be a much less confusing term than height or width.

Could we think in a similar way about the distance and time between two events? Is it possible that these terms too are terms of convenience, a result of observer perspective rather than intrinsic features of the events? In fact this is the case. It is possible to introduce a new characteristic that describes the separation of two events in space and time and does not depend on the state of an observer. This characteristic is often called "four-dimensional distance" or "4-duration." If we let D represent the coordinate separation between the two events and T the temporal separation, then 4-duration is defined as $\sqrt{D^2 - c^2 T^2}$. Thus, to measure the 4-duration between two events in your frame of reference, you measure the ordinary distance between them with a ruler (or by using position markers fixed in the frame) and the time between them as noted by clocks located at the events themselves, then calculate the difference between their squares and take its square root. It's a kind of Pythagorean theorem, in which the 4-duration is the hypotenuse and the time and spatial separation are the two sides. Instead of adding the squares, you subtract them. The beauty of 4-duration is that it is invariant. The 4-duration between two events is the same number to all observers frames if they express the distance and duration between the events in terms of their own coordinates.

The distinction between height and width is meaningful on Earth because of the presence of gravity. Gravity is a symmetry-breaking presence. Without the effect of gravity, as in the case of astronauts in a space capsule, there is no way to tell if you have been rotated from the horizontal to the vertical, so "up" and "down," "height" and "width" are only artifacts of a language developed on Earth. Relativity tells us that there is a similar case to be made for the words "spatial separation" and "temporal separation." It would seem that our own motion—that is, the motion of one inertial frame relative to another—"rotates" space into time, not unlike the way revolving our bodies would rotate height into width or up into down.

Why is it then that space and time seem distinct and different from each other, and why do they remain useful concepts to us? Is there something like gravity that breaks the symmetry between them? In fact, it is the slowness of our motions that makes space and time appear different to us and keeps them disconnected in our minds. This is not a real breaking of symmetry as in the case of gravity; it is simply that we limit ourselves to states of motion that tend to obscure the symmetry. It is more of a "bending" than breaking. In some distant future, when travel on rockets near the speed of light is commonplace, the distinction between space and time will become blurred. It is even possible that along with clocks and rulers, which we would probably still use in our own frames of reference (i.e., there would be a clock on the wall of the rocket ship so that all the passengers sharing that reference frame would know when lunch was being served), we will carry "4-duration meters" to help us maintain relations with friends and loved ones in other frames of reference.

Space and Time Become "Spacetime"

Three years after the publication of Einstein's revolutionary 1905 paper, the physicist Hermann Minkowski gave an address entitled "Space and Time" to the Assembly of German Natural Scientists and Physicians, in which he made the following remark: "Henceforth space by itself and time by itself are doomed to fade away into mere shadows, and only a kind of union of the two will preserve an independent reality."

With these introductory words, Minkowski launched into an explanation of Einstein's theory in which he introduced the concept of a four-dimensional "spacetime" to replace the previous concepts of a three-dimensional space and a one-dimensional time. Minkowski was carrying Einstein's theory to its logical conclusion. Since space and time have lost their independent meaning, we may as well start living in a true four-dimensional world.

Unfortunately, except for theoretical physicists, very few people believe we are living a four-dimensional existence. I'm not sure that theoretical physicists really think that way, either. The plain fact is that our brains simply can't deal with four dimensions, and since our lives haven't speeded up to near the speed of light, there is nothing about our existence that compels us to face up to a four-dimensional reality. Nevertheless, the concept of a four-dimensional spacetime, which is quite useful for interpreting special relativity, becomes absolutely critical to understanding Einstein's later theory of "general" relativity.

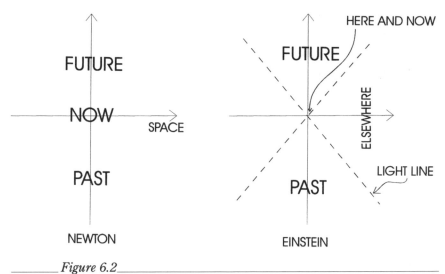

Figure 6.2
Newtonian and Einsteinian Universes

Picturing the Einsteinian Universe

It is worthwhile summing up what we have learned from Einstein by drawing two diagrams, one entitled "The Inertial Observer's Newtonian Universe," the other "The Inertial Observer's Einsteinian Universe." The Newtonian universe is a single universe shared by all inertial observers. It is divided into three parts: past, now (or present), and future. Newton's past consists of all the "nows" that have already occurred. His future consists of all the nows that are yet to be. His now is a three-dimensional universe shared by all inertial observers. Newton, it must be remembered, could not rule out some signaling mechanism that traveled at an infinite speed. He realized that the speed of light was finite but had no way of knowing that no other signal could exceed the speed of light. On the assumption that such an infinite signaling mechanism exists, we could all share a single now. Simultaneity would be absolute in the Newtonian universe.

The Einsteinian universe is quite different from Newton's. Einstein knows that no signal can travel faster than light and that light travels at the same speed for all inertial observers. Thus Einstein's universe is drawn for one observer only. There is in fact one universe for each observer, and in it the observer can describe distances and durations between events using an array of clocks and rulers. An inertial observer in a "different" universe will observe the same events occurring, but the times and distances between them will be represented by different numbers from those obtained by other observers. Each individualized universe is divided into four parts: past, future, here and now, and elsewhere. The first three are separated from elsewhere by intersecting lines called "light lines."

Newton's "now" encompassed all of the three-dimensional universe, because it would take no time at all to observe an infinitely fast signal that emanated from any event in that universe. Thus you could really "see" all of the stars "now" by using that signal (which wouldn't be light). Furthermore, all inertial observers would agree on that single now, because an infinite-speed signal would not have the amazing properties of light but would be infinite for all observers. While it is true that the signal would have the same infinite speed regardless of an observer's motion, this does not have the same consequences as a finite speed that has the same value to all observers. Einstein's "now," for a single inertial observer, can be only the unique position where the observer actually is. We shall see why this is so.

Suppose our observer, stationed at a particular here and now, takes note of an event. The fact that the event is seen means that light emanating from it reached the observer at the here and now. This, however, means that the event itself must have happened some time in our observer's past. In fact, any event that an observer sees, feels, or in any way senses (even by receiving a letter or telegram) must have occurred in the observer's past, somewhere on or between the two intersecting light lines. Only those events that are actually seen lie on the light lines themselves.

What about an event that occurs at some time and distance from the observer, such that no signal from the event, even one traveling at the speed of light, could possibly reach the observer at the particular here and now we are considering? Such an event is said to be "elsewhere." It does not lie within the past of the observer. In a similar vein, the future of the observer consists of all events that the observer, from the particular here and now, could possibly influence. The light lines play the role of boundaries, separating those things that could conceivably affect us or that we could conceivably affect from all things that are beyond our influence. There is no metaphysical "now" in this universe, in the sense that the observer can't ask, "I wonder what's happening 'now' in my friend's house?" (For some reason, I always think of the old song "I Wonder Who's Kissing Her Now?" when I discuss this issue.) That event in the observer's friend's house, whatever it may be, is in the observer's elsewhere. ("I Wonder Who's Kissing Her Elsewhere?" is much more appropriate, relativistically speaking.) Another observer moving relative to this one will not agree on the simultaneity of events, so "now" has also lost its absolute meaning to different observers.

Perhaps the best way to realize the strangeness of the concept of "now" is to go outside some clear evening and gaze up at the sky. Everything that you see is in your past (actually, on those light "boundaries"). Your eyes are here-and-now catching beams of light that were sent out millions of years ago, depending on just how far away the particular stars are. You are seeing each star as it was when it emitted that particular beam that your eyes are catching in the particular here and now.

In a true four-dimensional representation, the so-called light lines would not be lines at all; they would be imploding (from the past) and exploding (into the future) spheres of light. In a three-dimensional representation with space the horizontal axes and time the vertical, which is the most common textbook representation, the light lines become "light cones."

Relativistic Calculations

In the following section I will do a few "simple" calculations using the Lorentz transformations. I put the word "simple" in quotes because I have never found any of these calculations to be really simple, even when they are applied to the most elementary situations.

The Lorentz transformations provide a connection between the space and time coordinates of the same event as they would be determined by each of two inertial observers moving relative to each other. It is assumed that each observer has marked off a reference system with a set of spatial coordinates (one could have rulers nailed along the walls and floor) and has an arrangement of synchronized clocks positioned along the coordinate markers to determine a temporal coordinate. The clock arrangement is very important, because each observer needs a method of determining the time at which a distant event occurs. Remember, observers usually cannot be at the place of the event, so they must judge its time of occurrence remotely, determining what the clock closest to the event is reading. The ability to maintain the synchronization of clocks depends on an assumption that clocks will run at the same rate at all points in space-time. Actually, in general relativity, which deals with space-time where there is gravity, synchronization is impossible. In special relativity, however, space-time is empty, and clocks can indeed be synchronized.

There are several ways that synchronization can be achieved. The simplest would seem to be to gather all the clocks together at one place, set them all to the same time, and then carry them to their new positions. Einstein was so careful, however, that he worried about the possibility that moving the clocks might affect their settings. He suggested instead that the clocks be first positioned, then set. This could be done as follows. Set the clock that is positioned at the origin of coordinates (which I'll call the first clock) to some appropriate time, say exactly noon. Then go to the second clock, which I'll assume is one meter away, and look back at the first clock. As soon as you see 12:05 (any time will do) on the first clock, set the second clock to 12:05 plus 0.00000000333 seconds. Why the extra 0.00000000333 seconds? Because the first clock is actually reading that extra fraction of a second later by the time you see 12:05 on its face. It takes approximately 0.00000000333 seconds for the light carrying the image of the first clock to move to the second clock. In this way, an observer can move from clock to clock, setting each new one that same tiny bit ahead of the time seen on the previous clock. Having synchronized all clocks in this way, it is possible to say, "That event happened over there at noon!" The synchronized clocks have established a meaningful concept of time throughout the universe of the inertial observer.

Once inertial observers establish their coordinate markers and synchronized clocks, the Lorentz transformations have a meaning that can be agreed upon.

The Lorentz Transformations
Motion Assumed along *x*-Axis in Positive Direction

1. $x' = (x - vt) / (1 - \dfrac{v^2}{c^2})^{1/2}$

2. $y' = y$

3. $z' = z$

4. $t' = [t - (xv / c^2)] / (1 - \dfrac{v^2}{c^2})^{1/2}$

In these transformations the unprimed variables, x, y, z, and t, are the coordinates of an event measured by one inertial observer; the primed variables, x', y', z', and t', are the coordinates of the same event as measured by another inertial observer. Each has rulers and synchronized clocks set up in an inertial frame. The first, positioned at the origin of the primed system, is moving relative to the second (at the origin of the unprimed system) with the velocity v in the direction in which the x coordinate increases. Moreover, it is assumed that they were alongside each other when the clocks positioned at the origin of their coordinates read zero. Those are the ground rules for using the transformations in the form listed above.

The transformations can also be reversed, so that they give the unprimed variables in terms of the primed variables, by simply changing v to $-v$ and changing each primed variable to an unprimed one and vice versa. Changing v to $-v$ expresses the fact that if the unprimed observer sees the primed observer moving to the right (with a positive velocity), then the primed observer sees the unprimed one moving to the left (with a negative velocity). The v appearing in the transformation, being the velocity of one observer relative to the other, simply changes sign.

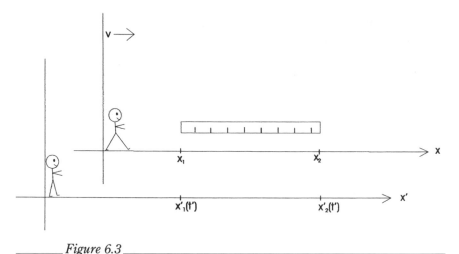

_____ *Figure 6.3* _____
Comparing the Length of a Meter Stick

Measuring Lengths

The simplest use of the transformations is to show that the length of a moving meter stick (a rigid rod one meter in length at rest) is less than a meter when measured by an observer with respect to whom it is moving. To use the transformations, we need two observers, and we have to convert everything to "events"—that is, points in four-dimensional space-time. Remember, the transformations don't provide lengths or durations directly; they simply relate the coordinates of observed events as determined by two relatively moving observers. It is up to us to choose the events so that a length or a duration can be deduced from them. Let us therefore assume that we are the observers relative to whom the meter stick is moving and that there is another observer, called the stationary observer, who is at rest relative to the stick. To set up the mathematics, I define a set of coordinates for the stationary observer and a set for us, the moving observers. Let me call the stationary observer's coordinates x, y, z, and t and our coordinates x', y', z', and t'. The stationary observer has already measured the stick and found it to be exactly one meter long.

Since we want to obtain the length of a meter stick, it seems reasonable for our two events to be the observation of the two ends of the stick. "I observe that one end of the stick is at point x'_1 on my coordinate axis at time t'_1 as read by my clock located at that same point" is one event. Similarly, "I observe that the other end of the stick is at point x'_2 at time t'_2" is another event. How can these two events be used to define a length, and how can we relate them to another pair of events by using the Lorentz transformations? Since we want to measure the meter stick, we have to observe both its ends at the same time in our reference system. Only then can we simply subtract x_1' from x_2' and call the result a length. In other words, the two events we need are, "I observe one end of the meter stick at x_1' and t'" and "I observe the other end of the meter stick at x_2' and also at t'." Only in terms of these simultaneous events will the length of the stick be simply $x_2' - x_1'$. Now we use the Lorentz transformations to deduce the position and time coordinates the stationary observer assigns to these same events.

To the observer at rest relative to the meter stick, the positions x_1' and x_2' of the two events in my frame are related to the positions x_1 and x_2 of the same two events by the transformations (given in the first equation):

$$x_1 = \frac{x_1' - vt'}{\sqrt{1 - \frac{v^2}{c^2}}}; \quad x_2 = \frac{x_2' - vt'}{\sqrt{1 - \frac{v^2}{c^2}}}.$$

Using these two expressions, I simply subtract x_1 from x_2 and cross multiply to obtain

$$x_2' - x_1' = (x_2 - x_1)\,(1 - \frac{v^2}{c^2})^{1/2}.$$

This result shows that the length of the moving meter stick measured by me, $x_2' - x_1'$, is less than the length of the stick as measured by the stationary observer, $x_2 - x_1$, by the factor $(1 - \frac{v^2}{c^2})^{1/2}$. This factor is always less than one. Note that the times, t_1 and t_2, at which the stationary observer sees me make my two observations will not be simultaneous to that person. (You must look at the appropriate Lorentz transformations to see this.) So why, you may ask, is $x_2 - x_1$ the length of the meter stick to the stationary observer if they represent nonsimultaneous position measurements? The answer is that the meter stick is not moving relative to the stationary observer, so it is irrelevant that the positions of its ends are observed at different times.

Could I have done this calculation in reverse order? Could I have related the results of a length measurement performed by the stationary observer to its value to an observer in the moving frame? In other words, couldn't I assume that the two events are the stationary observer locating end x_1 at t and end x_2 at t? Then, couldn't I use the Lorentz transformation to find x_1' and x_2'? Yes I could, but the values of the primed coordinates I would obtain that way would not correspond to observations done simultaneously by me. In other words, their difference would not be a length in the moving frame.

Measuring Time

Now that we have seen the length of a moving object reduced, we should be convinced that the elapsed time between two events also depends on the frame of reference from which those events are observed. In particular, I will take the events to be two readings of a single moving clock, so that the elapsed time between the events will be directly related to the rate of the clock. For this demonstration, I will imagine that there is a clock fixed at the origin of some frame of reference, which I'll call the rest frame of the clock. Let event 1 be its indication of some time I'll call 0, and let event 2 be its indication of some arbitrary later time t.

There is a second observer in a frame of reference relative to which the clock is moving with a velocity v in the positive x direction. I'll call that the moving-clock frame. This observer sees the clock fly past with a velocity v, but the clock happens to be alongside initially, so that event 1 also occurs at the second observer's origin of coordinates at the same time 0. But to this observer, event 2 occurs a time t' later at a positive distance vt' from the origin and is recorded on the clock that happens to be located at that position. Since all the clocks in the frame are synchronized, the difference in the two clock readings, even though they are at different places, will still be called the "elapsed time" by the observer. Our task is to determine t', the time indicated on the second clock, which therefore is the elapsed time between the two events in this moving-clock reference frame. It will not be the same time t that elapses in the clock's own rest frame.

To use the Lorentz transformations, we must first list the coordinates of the two events in each frame and decide which ones are unknown. For event 1, the coordinates in the rest frame are $x_1 = 0$ and $t_1 = 0$. The coordinates of this same event in the moving-clock frame are $x'_1 = 0$ and $t'_1 = 0$. Note that both observers are beside their respective clocks at this event. For event 2, the rest frame spatial coordinate is still the origin, so $x_2 = 0$, but the clock now reads t_2 seconds. In the moving-clock frame, the spatial coordinate of event 2 is $x'_2 = vt'_2$ (the moving clock having moved to the right with velocity v), and the stationary clock at that position reads t'_2 seconds.

We have to relate t'_2, the time of event 2 in the moving-clock frame, to t_2, the elapsed time in the rest frame. We do this by using the Lorentz transformation of the time coordinate:

$$t' = \frac{t + \dfrac{xv}{c^2}}{\sqrt{1 - \dfrac{v^2}{c^2}}}.$$

Since $x = 0$ at event 2,

$$t_2' = \frac{t_2}{\sqrt{1 - \dfrac{v^2}{c^2}}}.$$

Thus we have arrived at the result that t_2', the time that has elapsed between the two events as measured by clocks in the moving-clock frame, is greater than the corresponding elapsed time in the rest frame of the clock. I emphasize the fact that the elapsed time on the single clock in its rest frame is actually read on two different clocks in the moving-clock frame. However, since all the clocks in the moving-clock frame are synchronized, that observer would correctly say that t_2' seconds have elapsed on the clock at the origin. Since the two events are the readings of a single moving clock, this is the origin of the expression that moving clocks run slow. Note also that the particularly simple form

$$t_2' = \frac{t_2}{\sqrt{1 - \dfrac{v^2}{c^2}}}$$

is due to the fact that the clock in its rest frame remained at a single point, the origin. In general, the time difference between events that are separated in each of two relatively moving reference frames will have a more complicated form that includes the spatial separation.

Calculations Using Spacetime Distance

Although the Lorentz transformations are in a sense failure proof, using them requires great care choosing the correct events. Fortunately, many relativistic calculations can be carried out without ever writing down a single Lorentz transformation. An approach that avoids using these transformations involves the invariance of the four-dimensional distance between events. That's harder to say than it is to do. I've already introduced this concept earlier in the chapter. The four-dimensional distance between two events is analogous to the three-dimensional distance between two points in space. The difference is that whereas three-dimensional distance is obtained by summing three squares (Pythagorean theorem), four-dimensional length is the difference between one sum of squares and another single square.

A Trip to a Distant Star

One of the strangest results of special relativity is the fact that an interstellar voyager can survive a trip to a distant star that should take longer than a lifetime to complete, if the person travels quickly enough. Let's examine this process using the mathematics we have just learned and see how it occurs.

I will assume that a spaceship leaves Earth with an observer named "Voyager" aboard, while a close friend, named "Earthbound," remains at home. Neglecting very brief periods of acceleration and deceleration at takeoff and landing, we will have the spaceship travel at a constant speed of $0.9c$ (nine-tenths the speed of light) to an inhabitable planet circling a distant star that, by Earthbound's measurement, is exactly 100 light-years away. At the ship's given speed, the trip should take more than 100 years relative to Earthbound. Earthbound therefore does not expect Voyager to live long enough to complete the journey. However, if we compute how much time elapses on Voyager's clock, we shall get an entirely different result.

Using the concept of the equality of space-time distances between events for all inertial observers, we will need two events for the calculation. Let event 1 be the spaceship departing from Earth and event 2 be its arrival at the distant planet. These two events will be characterized by different space and time coordinates in the Earth system and the rocket system, but the spacetime distance between them will be the same in both systems.

Event 1 can be given simple coordinates in both systems by having both Earthbound and Voyager set their clocks to 0:00 as the rocket departs. Using unprimed coordinates, Earthbound calls the position of the rocket $x_1 = 0$, since the rocket is right on Earth, the origin of Earthbound's coordinates. Using primed coordinates, Voyager calls the position $x'_1 = 0$, since Earth is right at the rocketship, the origin of Voyager's coordinates. Event 1 thus occurs precisely at the positions of both participants, which are the origins of their respective coordinate systems.

Event 2, on the other hand, is described by different coordinates. To Earthbound, it occurs at the distant planet, so it has the position $x_2 = 100$ light-years. Correspondingly, it occurs at the time $t_2 = 111$ years, which is the time it takes for the rocket, moving at 0.9 times the speed of light, to travel 100 light-years in Earthbound's reference frame. Earthbound's clock will read 111 years when the spaceship reaches its destination. It will also be the simultaneous (to Earthbound) reading on a clock on the planet, since all the clocks in Earthbound's frame of reference read the same time at each instant. Note that if Earthbound could live as long as Methuselah and was actually looking at the clock on the distant planet with a powerful telescope when Voyager arrived, there would be a reading of 111 years on that clock but 211 years on the Earth clock. But this is immaterial to our calculations. All the clock readings that we use register on the clocks positioned at the events being described.

To Voyager, event 2 has the position $x_2' = 0$, since the distant planet is located right at the rocket ship when it arrives. To Voyager, of course, it is the planet that arrives at the rocket ship. Both events have exactly the same position coordinate to Voyager, even though they are widely separated to Earthbound.

What is the time coordinate of event 2 to Voyager? Since we do not know from any simple facts, we will simply call it t_2' and calculate it.

Using s to represent space-time distance (4-duration) and subscripts 1 and 2 to indicate the two events, the square of the space-time distance between events 1 and 2 is given by

$$s^2 = c^2 (t_2 - t_1)^2 - (x_2 - x_1)^2 = c^2 (t_2' - t_1')^2 - (x_2' - x_1')^2.$$

We have expressed s in both primed and unprimed coordinates in identical form. We can now insert the correct numerical and symbolic values:

$$(x_2 - x_1) = 100 \text{ light-years};$$

$$(t_2 - t_1) = 111 \text{ years};$$

$$(x_2' - x_1') = 0 \text{ light-years};$$

which gives

$$c^2(t_2'' - t_1')^2 = c^2(111 \text{ years})^2 - (100 \text{ light-years})^2.$$

Since $c = 1$ light-year per year (note the strange velocity units I am using), all the terms can be expressed in years, and the equation is easily solved:

$$(t_2' - t_1') = 48 \text{ years}.$$

Voyager is therefore only 48 years older by the elapsed time on the spaceship clock when arriving at the distant star. Voyager will be quite astonished, no doubt, to learn that the clock on the planet reads 111 years, for a trip that has taken only 48 years. We will shortly discuss this and other apparent discrepancies.

Another way of looking at this trip is in terms of length contractions and time dilations. If Voyager used a colossal ruler to measure the distance between Earth and the distant star as they moved past him at 0.9 c, the measurement would not yield the value 100 light-years as claimed by Earthbound, but a contracted value of 43.6 light-years, because the factor

$$\sqrt{(1 - \frac{v^2}{c^2})}$$

turns out to be 0.436. To Voyager, of course, it is Earth and distant star that are moving past at 0.9c, rather than the other way around. Voyager therefore

A

Image of t = 0   Unset

t=L/c

Sees t = 0; sets t = L/c

Clock being synchronized by a stationary observer

B

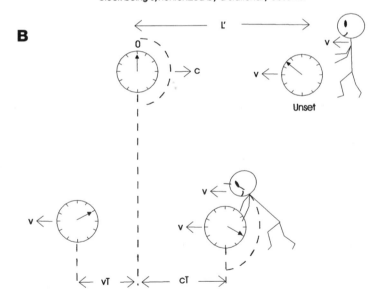

Synchronization seen by second observer moving to the right with speed v

Figure 6.4

Synchronizing Clocks

computes the time it should take for something that is 43.6 light-years long to move past at a speed of 0.9c and gets the answer 48 years. This is perfectly consistent with Voyager's clock reading. So Voyager explains the shorter than expected trip by noting that the distance traveled was less than originally reported by Earthbound, no doubt the result of faulty astronomical observations.

Earthbound, on the other hand, explains Voyager's unexpectedly minimal aging by the fact that all of Voyager's "clocks," including heart and biorhythms, are running slow. The time dilation factor, which is simply the inverse of the length contraction factor, is 2.29 (1 / 0.436 = 2.29), meaning that 2.29 seconds elapse on Earthbound's clocks for every second that elapses on Voyager's clock. Earthbound might not understand the detailed physics and biology of what is happening, but it certainly explains how Voyager survives the trip.

Of course, the time dilation and length contraction factors are symmetrical, so Voyager sees Earthbound's clock, heart, and biorhythms also slowing down. At this point it is easy to generate seemingly paradoxical data. If Voyager sees Earthbound's clocks running slow during the entire 48-year trip, then, by Voyager's calculations, shouldn't Earthbound have aged only 48 / 2.29 or 21 years? Yet Earthbound, by his own reckoning, would age 111 years during Voyager's trip, and the clock on the distant planet that Voyager finally sees also reads 111 years. How can Earthbound's clocks show 111 years elapsed, while Voyager has every reason to believe that only 21 years should have elapsed on those same clocks? Here we must be very careful if we are to obtain a reasonable resolution of this apparent paradox.

The Lack of Synchronicity of Clocks to Different Observers

Without doubt, one of the most puzzling results of relativity is the fact that a string of clocks synchronized in one frame of reference will not be synchronized in another frame of reference moving relative to the first. This fact, which, like many others, can be attributed to the invariance of the speed of light, is the reason for the discrepancy between the 111-year reading of the clock on the distant planet and the 21 years that elapse on earth relative to Voyager. I shall now attempt to explain how this occurs by first repeating the method by which an observer synchronizes two clocks that are a distance L apart. Then I will show how a moving observer determines that the clocks have not in fact been synchronized.

Figure 6.4a shows a stationary observer standing a distance L (at $x = L$) from a clock at the origin, waiting to see it register "0:00." The image of the origin clock arrives at L when the clock is already reading the time L/c, so the observer sets the clock at L to also read L/c. The two clocks are therefore synchronized.

In Figure 6.4b, a second observer, moving with velocity v with respect to the first one, is observing the synchronization process. According to this moving observer, the process results in a difference between the two clocks. The reasons are as follows. First of all, the moving observer sees the stationary observer moving toward the light that carries the image from the clock at the origin. This motion brings the image to that observer's eye in a time T, which is obtained by solving the simple algebraic equation

$$\text{(1)} \qquad cT + vT = L'.$$

In this equation, cT is the distance the light moves to reach the observer's eye, vT is the distance the observer has moved toward the light, and L' is the distance between the two clocks as measured by the moving observer. L' is not the same as L, since all distances are contracted because of the relative motion. Thus

$$\text{(2)} \qquad L' = L\sqrt{(1 - \frac{v^2}{c^2})}.$$

Inserting this value into the equation for T, we obtain

$$\text{(3)} \qquad T = \frac{1}{c+v} L\sqrt{(1 - \frac{v^2}{c^2})}.$$

This is what the moving observer expects the stationary observer to set his clock to. To make matters worse, during this time the origin clock has been running slow (as moving clocks do), so it is reading

$$\text{(4)} \qquad T\sqrt{(1 - \frac{v^2}{c^2})}$$

when the stationary observer sets the second clock to L/c. Therefore, instead of the two clocks reading the same time, they differ by

$$\text{(5)} \qquad \Delta T = \frac{L}{c} - T\sqrt{(1 - \frac{v^2}{c^2})},$$

which is the difference between the time just set on the second clock and the time simultaneously being registered on the origin clock relative to the moving observer. Replacing T by its value in equation (3) gives

$$\text{(6)} \qquad \Delta T = \frac{L}{c} - (\frac{L}{c+v})(1 - \frac{v^2}{c^2}).$$

A bit of algebraic manipulation produces the final answer:

(7)
$$\Delta T = \frac{Lv}{c^2}.$$

This simple expression represents the degree to which one observer's synchronized clocks are out of synchronization to another observer.

We can now see how Voyager explains the discrepancy between the time on the clock on the distant planet and the amount of time he knows has elapsed for Earthbound, who has remained at home. Recall that the clock on the distant planet reads 111 years, while Voyager knows that Earthbound has aged only 21 years. But Voyager also knows that Earthbound made a serious error when he synchronized his clocks. The clock on the distant planet is actually set too late by the amount

$$\Delta T = \frac{Lv}{c^2}.$$

Since L = 100 light-years and v = 0.9c, we conclude that ΔT = 90 years. Sure enough, if you subtract this from 111 years, you get the correct 21 years that Voyager expects to see on the clock that stayed on Earth.

The Twin Paradox

Probably the most famous of all paradoxes in special relativity is the so-called twin paradox, which is an extension of the problem we have just discussed (with Earthbound and Voyager conceived of as twins). The paradox arises if we allow Voyager to return to Earth.

At the time and place last seen, Voyager had aged 48 years but firmly believed that Earthbound had aged only 21 years. We explained all that. Voyager is correct. Earthbound, on the other hand, had aged 111 years and believed that Voyager had aged only 48 years. Earthbound is also correct. Thus each observer believes that the other has aged less. This may be disturbing, but it is not a paradox. It is simply a result of the nature of space and time measurements made by different observers in relative motion who become widely separated in space and time.

If Voyager makes a second trip, however, and immediately returns to Earth, supposedly we will have a paradox, because now Voyager should have aged by 222 years according to Earthbound's grandchildren, whereas Voyager, by self-reckoning, will be "merely" 96 years older. We no longer have a result of widely separated space and time measurements, because now both observers are located at the same place and time. Moreover, why isn't Earthbound's trip, away from and back to Voyager, exactly the same trip as Voyager's, away from and back to Earthbound? In short, since the trips are perfectly symmetrical, why doesn't special relativity give a single, noncontradictory answer for both of them?

This little puzzle really has two aspects to it, each of which requires a different explanation. The first aspect relates to the fact that we can't blame it all on special relativity. Two observers, each in inertial reference systems and in relative motion, can make seemingly contradictory observations, which are part and parcel of special relativity. The contradictions are explicable only while the observers remain in their states of motion. If the trips are truly symmetrical, however, we can't have the contradictions remain once they come to rest relative to each other. The explanation lies in the fact that Voyager, who must return to Earth, has to jump from one inertial frame of reference (outbound) to another (inbound), and is therefore not a single inertial observer for the entire trip, whereas Earthbound is. Thus we should not be shocked that the final ages of Earthbound and Voyager are different when they meet. The initial symmetry of their motions, when they were both inertial observers, has been broken.

Now we have explained one part of the paradox. A difference in aging is acceptable under the guidelines of special relativity, because the motion of the two observers is not symmetrical. We have yet to explain why the difference occurs and which of the two, Earthbound or Voyager, has actually aged more.

It is Earthbound who ages the extra 126 years. There is no question that Voyager's trip requires 222 years of Earth time. This is an incontrovertible result. It is an observation made by Earthbound's grandchildren, all of whom remain inertial observers. It is equally correct that Voyager ages only 48 years on each leg of the trip. Voyager too is an inertial observer on both the outward and return trips (albeit in different inertial frames), and it is incontrovertible that the distance Voyager is traveling at $0.9c$ is only 46 light years. How does Voyager explain the fact that the clock on Earth is reading an elapsed time of 222 years, while the clock in the spaceship is reading 96 years? The answer is quite simple: it is lack of proper synchronization on both legs of the trip. The clock on the distant planet, after all, was off by 90 years. Heading back, Voyager finds the clock on Earth also off by 90 years, although Voyager is probably at a loss to explain how it got that way. Moreover, the clocks all ran slow during both segments of the trip. By the way, we had better give Voyager a new clock for the trip back, since the rapid acceleration required at the turnaround would probably have damaged the original clock.

Seven

General Relativity

Special relativity firmly established the equivalence of all inertial reference frames. Unlike the equivalence originally postulated by Newton, Einstein's equivalence was a true symmetry of nature. It was an equivalence of inertial frames with respect to all the laws of physics, not just the laws of mechanics.

According to special relativity, the length of an object and the time that elapses between successive ticks of a clock, as measured by different inertial observers, are relative quantities. Each observer measures particular values that are correct for that observer. The mathematical formulation of the laws of physics and the speed of light, on the other hand, are not relative, but the same for all inertial observers.

There are many nonscientific concepts that we accept as being relative. Beauty, for example, is said to be "in the eye of the beholder," which is an elegant way of saying that it is relative to the observer. We may argue with our friends over whom we consider to be attractive, but ultimately the argument stops and we chalk it up to opinion. It seems that we now have to accept the fact that length and duration are also in the eye of the beholder, which appears to be far more difficult to accept than the similar statement about beauty.

Extending the Symmetry

Letters Einstein wrote to his friends show that he was not completely satisfied with special relativity. The more he thought about the fundamental importance of the laws of physics (as he had already made quite evident by declaring their invariance in special relativity), the more he believed that those laws should be the same for all observers, not just for those privileged to be in inertial reference frames. Laws of physics should not be able to distinguish your reference frame, however it is moving, from the frames of other observers. Einstein set out to prove the correctness of his intuition.

If you accept the symmetry of special relativity, then you believe that no natural phenomenon is sensitive to the velocity of the frame of reference in which it occurs or from which it is observed. However, if you are not willing to extend the symmetry so that it also applies to noninertial (i.e., accelerated) frames of reference, then you must believe that some phenomena are sensitive to the acceleration of the frame of reference in which they occur or from which they are observed. Does such a sensitivity actually exist? Is there something inherent in accelerated motion, but not in uniform motion, that makes itself felt and is observable? Although Einstein was convinced that the answer to this question should be "No," it seemed to be "Yes." Appearances notwithstanding, Einstein still believed that it was too restrictive to limit equivalence to inertial frames only; all frames should be equivalent, inertial and noninertial alike.

Einstein realized that symmetries didn't come into being just because one believed in them. In fact, there were very good reasons for believing that not all reference frames were equivalent. An observer in a noninertial reference frame, such as one that is rotating or accelerating in a straight line, notices a special kind of behavior not present in an inertial reference frame. There simply does not appear to be a symmetry with respect to noninertial frames. Let's examine this point in greater detail to see why it is true.

Is Acceleration Absolute?

If I should say that I am tall, smart, and good-looking, you would have every right to respond, "Compared to whom?" Most descriptive terms in our language are relative: they mean something only when compared to some arbitrary but usually agreed upon standard. If I say that I am moving fast, you also have the right to respond, "With respect to whom?" since speed is also a relative quantity. That is one of the results of special relativity. This means that the particular "whom" I am moving fast relative to has a right to claim to be moving fast relative to me. Relative speed is a two-edged sword; it gives two people the right to make the same statement.

Suppose I say that I am accelerating. Is this also the kind of statement to which you can respond, "With respect to whom?" If it turns out that the answer is "Yes," that acceleration is relative, then we should be able to discover a symmetry of nature similar to special relativity. Unfortunately, the answer seems to be "No," that acceleration is absolute, which is to say that it has a meaning even in the absence of other observers.

Suppose two automobiles are initially at rest alongside each other, but suddenly one of them moves away. If acceleration were relative, the driver of each automobile would have an equal right to claim to be accelerating with respect to the other. But can such a claim be verified—that is, tested by physical experiments so that the results that hold true for one car would also hold true for the other? This is exactly what special relativity assures us will happen for cars that move at constant speeds relative to each other. But where acceleration is concerned, the claim can be disproved, because it is easy to devise an experiment that will have a different outcome for the accelerating car from the stationary one. Simply have the driver of each car hang a pendulum from its roof. The pendulum in the stationary car will hang vertically downward, while the one in the accelerating car will hang at an angle to the vertical. Physics tells us that one car is accelerating, while the other car is not. Even though each driver sees the other car accelerating away in a symmetrical manner, nature imposes a difference that can be detected by the use of a pendulum. In fact, the driver's back will be as informative as the pendulum. Anyone who has ever sat in an accelerating car knows that you feel as though you were being pushed forcefully into the seat. You are certainly not forced back into your seat when you simply watch another car accelerate.

Noninertial Forces

What causes the pendulum to behave differently in one car from the other? What makes it possible to say of one car, "This is the one that is accelerating"? To the driver of the accelerating car, the pendulum appears to be pushed backward by a force. Forces experienced by observers in accelerating reference frames are called "noninertial" forces, just as the frames themselves are termed noninertial frames.

A rotating reference frame, such as a carousel or turntable, is also a noninertial frame. An observer standing on the rotating carousel and another observer standing on the ground alongside the carousel might each think the other was moving in a circular path and therefore accelerating. But nature will not be fooled. It is only the observer on the carousel who feels the presence of noninertial forces. A pendulum is as useful on the carousel for detecting its centripetal acceleration as it was in the car, where it determined the presence of linear acceleration. If each observer holds a pendulum, the one held by the observer on the carousel will swing outward, while the one held by the observer on the ground will hang straight down. The observer on the ground will not experience the force that tugs outward at the pendulum of the carousel observer, even though that observer is rotating relative to the carousel. This force, by the way, is called "centrifugal" force. If two observers, both stationed on a rotating carousel, throw a ball between themselves, they will observe the ball moving in a curved path, even though they throw it directly at each other. Well acquainted with Newton's first law, which says that objects move in straight lines unless acted on by forces, the observers will have to conclude that a force is acting on the ball. This force is called the "Coriolis" force. An observer standing on the ground sees no curved path whatsoever; rather this observer sees the carousel rotating beneath the ball, which always moves in a straight line. In short, centrifugal and Coriolis forces exist only in rotating reference frames, but are real to the observers who experience them.

Newton worried about the effects of rotating reference frames and the way nature knows which frame is doing the rotating. In his revolutionary book *Philosophiae Naturalis Principia Mathematica*, published in 1686, Newton described how he had determined the absolute nature of rotational motion by performing a simple experiment with a bucket of water. He suspended the bucket from the ceiling of his laboratory with a long cord. Twisting the cord and releasing it caused both the bucket and the water to spin rapidly, as though they were in a rotating frame of reference. Newton noticed that there was a brief initial period of time during which the bucket spun but the water did not. It took a few moments for the bucket's motion to be communicated to the water. Once they were both spinning, however, the water began to climb the side of the bucket, leaving its surface visibly depressed in the center. During the brief period before the water began to spin but while it was rotating rapidly with respect to the bucket, its surface remained level. Only when the water began rotating along with the bucket did its surface become sloped to the center. It was clearly not relative acceleration that moved the water; it was absolute acceleration.

Newton wanted to understand why the rotating motion caused an outward pull on the water. At first he thought that it might be the relative motion of the material of the bucket affecting the water and causing the phenomenon. If this were the case, you would expect a noticeable effect at the beginning of the experiment, when the bucket was spinning madly but the water had not yet begun to move. It is at this time that there is the greatest degree of relative motion between bucket and water. After repeating the experiment with buckets of various shapes and materials, Newton concluded that the effect was independent of the nature of the bucket. The relative motion between bucket and water produced no effect whatsoever; it was only when the water itself began to spin that something happened to it. The effect was clearly a function of the rate at which the water spun.

But what could possibly be connecting the spinning of the water to the outward force on it? If the bucket itself was not tugging on the water, the only other thing that could possibly be affecting the water was the very space itself in which the water spun. Newton was forced to conclude that space itself provides the mechanism through which rotational motion produces an outward force. It is not that the water is rotating *relative to* something; it is that the water is rotating *in* something! Somehow space must be of sufficient substance to affect rotating matter. Newton was unwilling to speculate further on the essence of space, but he called it "absolute," implying that an observer could always detect noninertial motion with respect to it.

The great German philosopher Gottfried Wilhelm von Leibniz was a contemporary of Newton who disagreed with him on most issues. The two were very different sorts of people who were neither personally nor intellectually on the best of terms. Leibniz challenged Newton's conclusion that space could be a mechanism by which states of motion become detectable. To Leibniz, space was not a "thing"; it was the absence of all things, a mental construction, a creation; it was what the mind invented to allow us a definition of "separation." It was patently ludicrous that a mental construction should produce physical effects. The argument on this point between Newton and Leibniz went on for many years and continued between their respective adherents long after they both died.

Good arguments, however, never die. In the late 1800s, Ernst Mach, the positivist scientist so respected by Einstein, entered the argument on the side of Leibniz. Mach's positivism could not allow the attribution of physical qualities to space, just as it could not accept the existence of an intangible ether. Mach, however, applied a more substantive idea than Leibniz's. He reasoned as follows. Suppose, instead of rotating the pail of water, one held the pail fixed and rotated all the matter in the rest of the universe. Since no observer could possibly detect any difference in these two sets of circumstances, the water in the pail would still have to rise. Yet rotating the universe would not affect the space in which the pail was positioned, so the water could not be climbing because of the effect of space on it. Mach therefore concluded that the only possible cause of the rising water was the rotation of all the stars and other matter in the universe relative to the pail.

Einstein called this interesting idea "Mach's principle." Physical effects on objects in noninertial reference frames were produced by the noninertial motion of those objects relative to all the rest of the matter in the universe. Indeed, Mach argued that if all the matter in the universe were removed, there would be no conceivable way in which motion could be even defined, let alone detected by the senses. Mach's principle had the satisfying property that only matter was affecting matter, an interaction that would have to be reciprocal. If it were space affecting matter, the interaction would be one-way only, because Mach believed that space is not affected by matter.

The implication of Mach's idea is that acceleration *is* relative. Newton's error in concluding otherwise was that he rotated only the pail relative to the water, whereas he should have rotated the entire universe. Had he done that, the stationary water would most assuredly have climbed up the side of the bucket. The reason your pendulum doesn't swing back when the other car accelerates is that it is only the other car. If all the matter in the universe were made to accelerate, then, according to Mach, your pendulum would take notice.

When Einstein confronted the same question that had faced Newton, Leibniz, and Mach, he discovered appealing elements in each of their arguments. Although, like Mach and Leibniz, he rejected the notion that a space without substance could produce measurable effects, he liked Newton's idea that the effects were produced by "something" that was actually there at the position of the bucket. He also liked Mach's notion that the stars and other matter in the universe were the ultimate cause of the noninertial effects, because an interaction between matter "here" and matter "there" would be two-way. Matter can affect matter. But Einstein also knew from special relativity that any interaction between a spinning bucket and all the stars in the universe, near and far, would take a great deal of time to occur. No influence between material objects could travel faster than the speed of light, so a bucket that was spinning "now" would not feel the effects of a distant star for many years. To Einstein, this kind of time delay made a theory of direct interaction among distant objects extremely unattractive. Einstein much preferred the approach that Maxwell had successfully used in describing electromagnetic phenomena, a "field theory."

The concept of a field was not new in physics. It had already replaced the older "action at a distance" that Newton used to describe the force of gravity between separated objects. A field can be defined as a condition of the region of space that surrounds any object, a condition that is capable of exerting a force (or other effect) on another object. Instead of thinking of an object as exerting a force directly on another object and only on that object, we imagine instead that the object produces a field everywhere in space, even at places where no other objects happen yet to be. It is the field that is then responsible for producing the force when another object is placed in it. In short, certain objects produce fields, and those fields subsequently exert forces on other objects placed in them. The field becomes an active entity that takes on a life of its own, so to speak. The equations that describe that life are called field equations. Maxwell's equations are field equations; that is, they describe the behavior of electric and magnetic fields. There is a completely separate equation that describes the forces these fields exert on other objects.

The field concept is enormously useful in devising physical theories. Action at a distance requires that the distance between objects be carefully monitored. If, for example, two objects are separated by a distance d and one object moves relative to the other, neither object can be affected by the motion of its companion until at least a time d/c (where c is the speed of light) has elapsed. Keeping track of time delays between moving objects is difficult, yet this is the sort of thing Mach's principle would have required. In a field theory, however, the field is the bearer of all messages between interacting objects. One object (called the "source") produces the field, and the other object is simply located in the field at some position in spacetime. All of the burden imposed by motion of the source is transferred to the field. I do not mean to imply that using fields make matters simple. Far from it! But it allows a systematic treatment of the interaction between objects that action at a distance makes much more difficult. Einstein felt that whatever caused the water to rise in Newton's bucket should be describable by a field as opposed to action at a distance. This meant that the influence of the matter in the universe suggested by Mach would have to be related to some kind of field, and it would be the field that produced the noninertial effects.

Inertial Mass versus Gravitational Mass

Another issue bothering Einstein that was even more fundamental than Mach's principle was the relationship between inertial and gravitational mass. According to Newtonian mechanics, the motion of an object produced by the gravitational attraction between it and Earth involved two entirely different masses. Newton's second law, $F = m_i a$, contained the so-called inertial mass m_i, which was a measure of an object's resistance to being pushed or pulled by a force, F, of whatever origin. On the other hand, on the surface of Earth, gravity exerts a force on objects whose value is $m_g g$, where g is a constant (depending on certain properties of Earth) and m_g is the so-called gravitational mass of the object being acted on by Earth. This gravitational mass determines how strongly other masses pull on that particular object with their own gravitational forces. In a completely symmetrical manner, the gravitational mass also determines the strength of the gravitational force exerted by that object on other objects. There was no reason that an object's inertial mass, which measures its resistance to being accelerated by all forces, should be related to its gravitational mass, which determines only its susceptibility to gravitational attraction. The two masses played entirely different roles in the physical scheme of things.

When Newton's second law is used to determine the acceleration a of an object on which the external force $\boldsymbol{F}$ is Earth's pull of gravity, both masses come into play. By the definition of the gravitational force, $\boldsymbol{F} = m_g\boldsymbol{g}$. By Newton's second law, $\boldsymbol{F} = m_i\boldsymbol{a}$. By equating the F in each of these two equations, we obtain $m_g\boldsymbol{g} = m_i\boldsymbol{a}$. This in turn leads to the simple result $\boldsymbol{a} = (m_g/m_i)\boldsymbol{g}$. Thus the acceleration of an object acted on by thE earth's gravity at Earth's surface involves the ratio of the object's gravitational mass and inertial mass, which one would expect to be different for different objects. One would therefore expect that different objects should fall with different accelerations. Expectations notwithstanding, all objects are observed to fall under gravity with the same acceleration, g. This implies that the ratio of gravitational to inertial mass is the same for all objects, and in fact is numerically equal to 1. An experiment carried out by Roland von Eotvos in 1889 showed that the ratio of these masses does not differ from object to object by more than one part in 10^{-9}.

This unexpected relationship had been tested over and over again since the time of Galileo, and according to all experimental observations, the inertial and gravitational masses of all objects are equal. Is this some colossal coincidence? Are these two quantities actually different, but somehow numerically equal? Or is their numerical equality an indication of something much deeper, the fact that they are the same quantities? To Einstein, these were crucial questions. The reason that objects behave differently in noninertial reference frames from inertial frames, as when water rises in a spinning bucket, is precisely that they have inertial mass. This is the mass that makes objects resist being accelerated and move in straight lines with constant speed unless they are acted on by forces. In its simplest terms, Mach's principle attributes inertial mass to the presence of gravitational mass in the rest of the universe, whereas Newton was attributing it to the intrinsic nature of space. Whichever was correct, it would be very nice indeed if inertial mass could be eliminated as an independent property of matter and relegated to being an effect of something external. Since inertial mass breaks the symmetry between inertial and noninertial frames, eliminating it as an independent property of matter might somehow restore that symmetry. Gravitational mass, on the other hand, seemed to have nothing to do with the difference between inertial and noninertial reference frames.

The more Einstein thought about this, however, the more he realized that it was incorrect; gravitational mass had a great deal to do with reference frames. How does an observer recognize an inertial reference frame? By noting that an object moves in a straight line at constant speed when acted on by no net external force. But are there any regions in the universe that can truly be said to be free of forces? In short, can inertial reference frames even be found? Certainly one can eliminate all the obvious forces, such as those due to springs, ropes, levers, arms, legs, and the like, simply by not using any such objects. Electric and magnetic forces are trickier to do away with, but even they can be eliminated by means of shielding. This means you have to perform all your experiments inside a metal cage, which is uncomfortable, inconvenient, and expensive but completely doable. Finally, noninertial forces, like centrifugal and Coriolis forces, can be eliminated by jumping into a reference frame that moves with constant velocity, the frame that Newton called inertial. But gravitational forces cannot be eliminated. Unlike electromagnetic forces, they cannot be shielded, and you certainly can't turn them off.

It looks as though the search for a true inertial reference frame is a hopeless task. Unless, thought Einstein, gravitational mass and inertial mass were actually completely equivalent. If this were indeed the case, then gravitational forces could be eliminated simply by placing oneself in the proper noninertial reference frame, as we shall see in a moment. As matters stood, these two supposedly different masses were constant sources of irritation. Inertial mass, making itself felt whenever you jumped into a noninertial reference frame, destroyed a beautiful potential symmetry; gravitational mass made it difficult to even find a true inertial frame. Einstein was about to make a virtue out of necessity. He would restore the complete symmetry of all frames, not by eliminating inertial mass, but by declaring its complete equivalence to gravitational mass and demonstrating that, as a result, noninertial forces and gravitational forces are indistinguishable. This declaration of equivalence was as bold as Einstein's earlier declaration that the speed of light is the same for all observers. Once again, Einstein eliminated a previously believed distinction and, in so doing, affected the limits of possible knowledge.

In a life filled with brilliant insights, one of Einstein's most brilliant was his realization that the equality of inertial and gravitational masses caused an observer in a noninertial reference frame to feel the same sorts of effects as would be produced by a gravitational field. The observer would be unable to distinguish between noninertial and gravitational forces. Imagine an observer in an elevator car in a gravity-free region of outer space, being pulled upward with a constant acceleration g. This observer would experience all the same effects in the car on the surface of Earth. In the upwardly accelerated car, any object dropped by the observer, regardless of its inertial mass, would "fall" to the floor with the acceleration g. Of course, a second observer outside the car would see the floor of the car accelerating upward to meet a stationary object, but we must put ourselves in the place of the observer in the elevator. For us, there is no external reality, only the reality of the interior of the elevator. The object falls.

Place the same car on Earth's surface, and any object dropped by the observer would do exactly the same thing. This time, it is Earth's gravity that accelerates the falling object. But within the confines of the elevator car, the observer has no means of determining the difference between the two cases. To be perfectly correct, the car should be small in both horizontal and vertical dimensions. It should be narrow because gravity points toward the center of Earth, and this would cause a slight but noticeable deviation in the direction of its pull at opposite ends of a wide car. It should be short because the inverse square dependence of gravity on distance from Earth's center renders the force slightly greater at your feet than at your head.

No true gravitational field can be completely eliminated by accelerated motion. If you leap from a roof, your body will feel as though it had no weight. You have effectively eliminated the net pull of gravity. Your body will, however, feel ever so slightly stretched and warped by what are called gravity's "tidal forces," which are caused by the differences in the force of gravity between your head and feet. One must settle for the elimination of these tidal forces only to within a very high degree of precision in a very small region. To the same degree, an observer in an elevator car on Earth would lose the effects of gravity if the car were to fall freely down its shaft. In such a situation, any object released from the observer's hand would simply hang in space, as though gravity had been turned off. Thus the noninertial accelerating car can either produce all the effects of gravity when gravity is not there or cancel all the effects of gravity when it really is there. "Noninertialness" and gravity are indistinguishable (at least in a small region of space). Einstein called his creation the "principle of equivalence."

If you jump out of a plane and free-fall before opening your parachute, you will eliminate gravity. Take something out of your pocket and try to drop it while you are falling. It does not fall. Look at a videotape of astronauts circling Earth; they are said to be weightless, but actually they are in a state of free-fall. All the effects of gravity that you usually experience are a result of the fact that you insist on maintaining yourself in what you think is an inertial system (i.e., sitting in a chair in a nonmoving room). The weight you feel is due to your stubborn insistence on being inertial. Gravity can be thought of as the noninertial force you feel when you are in a reference frame that is not falling freely. In that sense, it is not different from the centrifugal and Coriolis forces experienced in a rotating frame.

We have just seen how to eliminate the effects of gravity. You can also do quite the opposite. You can create the effects of gravity where it does not already exist by placing yourself in the proper noninertial frame. If you stand on a scale in an upwardly accelerating elevator, you will notice that you weigh more. You have not put on any gravitational mass; you are simply experiencing the effects on your inertial mass of being in a noninertial reference frame (the elevator). The equivalence of inertial and gravitational mass will ensure that you cannot tell the difference.

Einstein extended his thoughts to their positivistic limit. If you cannot tell the difference between the effect of gravity, which is supposedly a "true" force, and the effects produced by noninertial frames, which would normally be considered illusory "pseudoforces" (because an "external" observer could see what was "really" going on), then the two effects are in fact one and the same and completely equivalent. This, as we have seen, is why Einstein called this amazing new concept the principle of equivalence. We can now propose an extended symmetry. In a universe devoid of gravitational forces or even in a region of the universe so far away from significant masses that gravity is insignificant, inertial reference frames can be easily defined by eliminating all nongravitational forces and testing for the applicability of Newton's first law. All the inertial reference frames discovered in this manner will be equivalent, because in them the laws of physics have the same form. This is the symmetry called special relativity, and the mathematical requirement that the equations take on exactly the same form is called Lorentz invariance. It will now be replaced by a much more general symmetry.

In a universe containing gravitational masses and energy, all reference frames are equivalent, and in them all the laws of physics take on the same general form. However, you have to relinquish the erroneous belief that you can tell the difference between gravity and noninertial effects. This new symmetry is called "general relativity."

As usual, the symmetry expresses a form of ignorance. "I might be in a noninertial reference frame, but then again I might be in a field of gravity. I simply can't tell." It seems remarkable that the addition of gravity to the universe allows us to extend the symmetry from inertial frames to all frames. It might be equally correct to say that gravity is present in the universe *because* of the symmetry among all reference frames. The price we pay to have all the laws of physics the same to all observers is the existence of a "force" that we have come to call gravity. We shall see that the principle of equivalence and the general theory of relativity that followed from it enabled Einstein to deduce a radically new explanation for the effects of gravity, an explanation that has the impact of a creation as opposed to a discovery.

A Special Noninertial Frame Becomes Inertial

Even though all reference frames are equivalent, the "first among equals" is the frame that falls freely in a gravitational field. It is the frame voluntarily chosen by the free-falling sky diver or involuntarily produced within a freely falling elevator. It is also the frame of astronauts orbiting Earth in a space station. In this special reference frame, (nearly) all effects of gravity are eliminated, and in all other respects it is at each instant of time inertial in the Newtonian, Einsteinian, and Minkowskian sense. In this reference frame, therefore, all the results of special relativity remain true. There is one warning that must be heeded, however. This gravity-eliminating, freely falling frame must be very small in spatial extent. Gravity can be eliminated only in a very small region about a freely falling observer. The appropriate jargon is that a freely falling observer eliminates gravity "locally" but not "globally."

In a universe containing gravitational masses, a freely falling reference frame can be thought of as a legitimate inertial reference frame, as can any frame that is instantaneously moving relative to it with a constant velocity. In this frame an object acted on by no forces moves in a straight line as though gravity were not present. Einstein removed gravity as an impediment to discovering inertial frames. In the general theory, an inertial reference frame is thus any frame in which the rules of special relativity hold true. Any observations made in such frames must be connected by the Lorentz transformations of special relativity. We shall see a bit farther on how this fact gives insights into the effects of gravitational fields.

How is the principle of equivalence related to Mach's principle or Newton's conception of absolute space? If the principle of equivalence is correct, then the noninertial forces pulling water up the sides of a spinning bucket can with equal correctness be interpreted as some type of gravitational force produced by the entire mass of the universe spinning around a stationary bucket. Can this be proved? Unfortunately, it would require a calculation of the gravitational field of an entire spinning universe, something that physicists are currently incapable of doing. As a corollary to this, imagine that the universe were made smaller by the elimination of some of its mass. It is reasonable to expect that this would reduce the effects of its spinning gravity. But that would mean that the noninertial forces on the bucket would also be reduced, implying that the inertial mass of the water was smaller. In short, we might have here a connection between the total mass of the universe and the mass of a single object in it. This is an aspect of Einstein's work that needs much more investigation, so we will leave it for now and return to something more mysterious but much better understood, the curvature of space-time.

Once Einstein had shown that the observational effects of being in noninertial reference frames were equivalent to those in gravitational fields, it remained for him to define the properties of such frames in greater detail. He had already studied the properties of inertial frames in special relativity and discovered that the relationship between spatial separations and temporal durations determined by observers in different frames was quite strange and counterintuitive. Surely something similar (and possibly even more counterintuitive) would occur for observers in noninertial frames.

In formulating the properties of inertial reference frames for special relativity, Einstein had gone to the heart of the concepts of space and time, demonstrating that properties usually associated with them could be unambiguously defined only in terms of the methods by which they are measured. When he examined these methods, Einstein discovered that length and duration had each lost their absolute nature. In general relativity, Einstein now launched a second fundamental attack on the nature of space and time. He would show that measurements of distance and duration made by an observer in a noninertial reference frame or a field of gravity can and should be interpreted as being produced by a "curvature" of four-dimensional spacetime. In other words, the behavior of clocks and rulers, both in gravity and noninertial systems, is such that the observer using them will be convinced that space-time has bizarre properties.

In the following sections, we shall examine in greater detail what is meant by a "curved space" or a "curved spacetime." We shall see that it is a concept intimately related to the way we do geometry and define distance and duration. On the physical or sensory level, we can think of gravity as being the result of a conflict between our insistence (which, conceivably, is largely genetic in origin) that spacetime is flat, whereas in fact it is curved. To maintain our illusion of normalcy, we "invent" a field of force called gravity. This is analogous to the way in which noninertial forces like centrifugal and Coriolis forces are imposed upon a rotating frame of reference. If we leave the rotating frame, we leave behind the centrifugal and Coriolis forces. If we leave the security of our rooms and jump headlong into a free-fall in space, we leave gravity behind.

As I mentioned earlier, because we live on a rotating planet, we are in a noninertial reference frame from birth to death. However, we do not think of Earth as rotating; we think of it as solid and unmoving. Because of that intentional and very useful self-delusion, when we notice that objects that should be moving in straight lines (neglecting for the moment their downward curvature due to gravity) actually move along curved paths, we insist that there must be certain forces acting on them. Of course, to an inertial observer out in space, these objects actually are moving in straight lines; it is we on Earth who are rotating out from beneath them. This makes it appear to us that they are traveling in curved paths. To the observer in space, there need be no additional forces at all. Centrifugal and Coriolis forces are *created* by earth-bound observers to maintain the fiction of an "inertial" earth. In the same way, gravity is created to maintain the fiction of a space-time that is "flat."

In the context of discussing the purely mathematical concepts of space curvature—that is, four-dimensional geometry in a curved space—I will show that the geometrical nature of space must be discovered by experiment, not established by belief or intuition. Finally, I will show how curvature of spacetime actually can produce the effects that we have come to call gravity.

Surfaces and Spaces, Flat and Curved

The geometry that we learn in high school is correct only on flat surfaces. It was invented by the Greek mathematician Euclid sometime in the third century B.C.E. In his honor, we call this geometry two-dimensional Euclidean geometry, or plane geometry. For the purposes of this book, the only feature of Euclidean geometry that I will be interested in is the nature of the distance between two points. I will initially use the term "distance" in its ordinary sense, what is measured by a ruler. Later on, as we expand our horizons and venture onto a curved surface, the distance between points will have to be measured by a succession of very small rulers. Finally, as we jump headfirst into a curved space or spacetime, the entire concept of distance will have to be rethought. Because time as well as space will be involved, we will have to use clocks as well as rulers in our measurements.

Distances and Coordinates

Nearly everything you could want to know about the nature of a surface can be discovered by studying the relationship of the distance between two neighboring points to the coordinate values that label them. Coordinates, as we have seen, are arbitrary labels that are placed on a surface to locate points on it unambiguously. Their sole task is to cover all or part of the surface so that every point in their jurisdiction has some label.

Certain kinds of coordinates are so easy to use that they have become universally accepted; but nothing compels you to follow the herd. Let me also point out that when I say that two points are neighboring, I mean that in a mathematical sense they are an infinitesimal or a "differential" distance apart. This kind of nearness is very important in dealing with curved surfaces, because it is easiest to say meaningful things about very small separations and then make deductions about larger distances by integration (adding up small effects).

There are about as many different kinds of coordinate systems as there are mathematicians who invent them. Most of the time, coordinates are constructed in some abstract way, in the imagination, so to speak, without regard to the

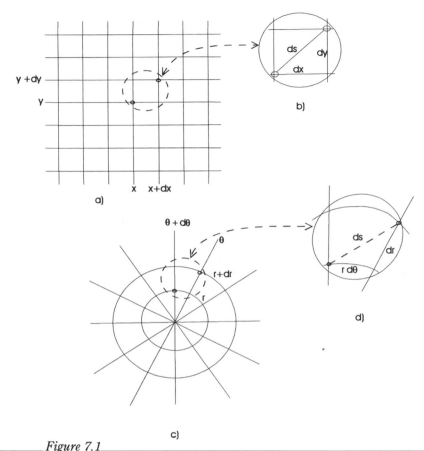

_____ *Figure 7.1* _____
Labeling Points in Cartesian and Polar Coordinates

shape of the surface that they will eventually label. The coordinates come first; the surface comes later. Sometimes, however, it is useful to draw coordinates directly on a surface, following whatever shape the surface happens to have. These coordinates may make it easy to describe positions on the surface, but considered apart from the surface, they will probably be extremely complicated. When distances are to be measured, however, they must be measured on the surface itself. A ruler has no choice but to follow the surface. The distances are measured along the surface; the coordinates are the numbers that happen to be on the surface where the distances are measured.

The simplest place to start is with a flat surface. If I mark two points on the surface, I can arrange them to be at the intersection of two nearby sets of coordinate lines, as shown in figure 7.1. The points can be labeled using either Cartesian coordinates (lines crossing at right angles) or polar coordinates (concentric circles and radial lines). With Cartesian coordinates, if the position of one point is labeled (x, y), then the position of the other can be labeled $(x + dx, y + dy)$, where dx and dy symbolize very small (mathematically, differential) distances measured along the coordinate lines. The shortest distance between the points as measured directly by a ruler, labeled ds, can now be expressed in terms of the distances dx and dy marked along the coordinate lines. I do this by applying the Pythagorean theorem to the tiny right triangle formed by ds (the hypotenuse), dx, and dy (fig. 7.1b). The result is

(1) $$(ds)^2 = (dx)^2 + (dy)^2.$$

Although it appears deceptively simple, this is quite an important result. It demonstrates that a flat surface can in fact be covered with Cartesian coordinates, which are defined as those that give this simple relationship between the shortest distance separating two points and the coordinate labels of those points. It turns out that this can be done only on a flat surface. You can't cover a globe or the saddle of a horse with Cartesian coordinates, no matter how hard you try.

A relation such as the one in equation (1), connecting infinitesimal distances with their corresponding coordinate differences, is called a "metric" (measuring) relationship. If I choose to cover my flat surface with a different set of coordinate labels, such as the polar coordinates shown in figure 7.1c, I will get a different metric relationship. With the polar coordinates, it is traditional to use (r, θ) to designate location. Every point is on some circle, which gives it a radial position r, and every point is on some radial line, which gives it an angular position θ. The distance between two nearby points is related to their radial and angular separations according to the relationship

(2) $$(ds)^2 = (dr)^2 + (rd\theta)^2.$$

This metric relationship for polar coordinates is quite different from the one using Cartesian coordinates, since the second term on the right involves both r and θ. If we chose to use still other coordinate systems than Cartesian or polar, we could produce even more complicated metric relationships. Nevertheless, the hallmark of a flat surface is the fact that Cartesian coordinates are possible and their metric relationship is the simple one given in formula (1).

Suppose we climb aboard a curved surface, like the surface of a sphere. The planet Earth comes to mind. There are now several different ways of setting up a coordinate system. If I happen to believe I am still on a flat surface, I could attempt to set up a system of polar coordinates by drawing radial lines from a center and surrounding the center with concentric circles. If I attempted to relate the distance between two nearby points to their coordinate differences, however, I would discover that I am not obtaining the metric relationship of equation (2). For example, the circumferences of neighboring concentric circles (a distance dr apart) do not differ by $2\pi(dr)$. In this way, I could determine the curved nature of my surface without ever leaving it. A little farther on in the book I will go through such an exercise as we investigate the curvature of a space rather than a surface.

What would happen if I attempted to draw Cartesian coordinates on the surface of a sphere? It is possible to mark off very small regions within which Cartesian coordinates would work. This is like covering the surface of a balloon with postage stamps and putting Cartesian coordinates on each of them. Each stamp is "flat" because it is so small, and therefore Cartesian coordinates can be drawn on it. The coordinates on neighboring stamps would not meet smoothly, however, so you wouldn't really have a single Cartesian system on the entire spherical surface. On Earth, you can certainly draw Cartesian coordinates on something as small as a football field; in fact that's exactly what the yard lines and sidelines are. If you are constructing the Alaskan pipeline, however, you will have to either curve the pipe noticeably or else connect straight segments with curved elbows. Mathematicians say that all curved surfaces are "locally flat" or "locally Euclidean," meaning that you can draw Cartesian coordinates in a very small region. However, they are not "globally flat," so you can't extend the individual sets of Cartesian coordinates to form a single coordinate system that covers the entire surface.

One of the useful features of a global coordinate system is the fact that coordinate differences don't change. By this I mean that the coordinate line $x = 1$ and the coordinate line $x = 2$ are always one coordinate unit apart, wherever they may be. Now this doesn't mean that they are one *distance* unit apart, as might be measured by a ruler. For example, the interstate highway system covers the United States in a global fashion and can be thought of as a coordinate system. The various numbered routes crisscross the country in a continuous fashion, with the odd numbers going north-south and the even numbers going east-west. This means that route 95 and route 93 are one route number apart. They will always be one route number apart; there will never be a route 93.5 that suddenly appears between them. Of course, the actual distance between them will differ greatly as they cross the country, but they will remain one route number apart. If you wanted to know the distance in miles between the intersection of routes 95 and 66 and routes 93 and 64, you would need some sort of formula relating route number differences to distances. This is precisely what the metric is. The locally Euclidean nature of curved surfaces is analogous to the locally gravitational-field-free region surrounding a freely falling observer. Doing geometry on a football field is like doing physics in a freely falling elevator.

The notion of a flat or curved two-dimensional surface can be extended to three dimensions to produce a flat or curved space. (Note: I will call anything with three or more dimensions a "space".) A three-dimensional space is therefore called flat if it can be labeled with Cartesian coordinates. In a flat, three-dimensional space, the Cartesian metric is

$$(3) \qquad (ds)^2 = (dx)^2 + (dy)^2 + (dz)^2.$$

You might wonder how these labels can be drawn in space, where there seems to be nothing to draw on. Think in terms of a three-dimensional gridwork that fills space, or else imagine that space is a solid block of foam rubber on which you can write with a pen or pencil. In any event, if you can install a single set of Cartesian coordinates in the three-dimensional space, the space is said to be flat and to have a Euclidean geometry.

Curved Space

What does it mean to say that space is curved? Intuitively, it is much more difficult to visualize a curved space than a curved surface. After all, each of us has seen and touched curved surfaces, but a curved space is another matter entirely. Fortunately you don't have to worry about visualizing a curved space if you simply use the concepts we have been discussing. A curved space is one that does not have a Euclidean geometry, because a Cartesian coordinate system cannot be constructed (except locally). It may not even be curved in the sense that a surface curves; it may actually be stretched or contracted.

Forget about trying to look at curved space; just draw coordinates and express infinitesimal distances in terms of them. If you get the Cartesian result, space is flat; if you don't, it isn't. The question of whether or not you are moving around in a curved space will not be answered by simply looking at that space. It will be answered by careful measurements.

Four-Dimensional Spacetime and Its Geometry

When Einstein completed his work on special relativity, it quickly became apparent that a four-dimensional spacetime would be a very convenient setting in which to apply his results. Now it turns out that the geometry of spacetime, as conceived by Minkowski, is not at all the same as Euclidean geometry, because the Pythagorean theorem does not exist in its usual form. Four-dimensional spacetime is an entirely new creation; Pythagoras, in his wildest dreams, never would have imagined it. It will be our task, as it was Minkowski's before us, to find the correct relationship between the "hypotenuse" and the "legs" of a four-dimensional spacetime triangle. To do this, we will have to use the properties of light along with the properties of ordinary distance, because lines drawn in spacetime extend into both time and space.

As we have seen, distances in spacetime are distances between events, which implies a spatial as well as a temporal aspect to this particular kind of distance. Minkowski defined the relationship between a small spacetime distance ds and the small corresponding coordinate changes dx, dy, dz, and dt to be

(4) $$(ds)^2 = (cdt)^2 - (dx)^2 - (dy)^2 - (dz)^2.$$

In this expression, dx, dy, and dz are the familiar Cartesian coordinate distances between the two events. The term cdt is unusual, however. It represents the spatial distance that light would move during the amount of time, dt, that separates the events. These four distances are now combined in a way that is only vaguely like the familiar Pythagorean relationship between distances, which would have been a *sum* of squares. But Minkowski didn't simply want to copy Pythagoras; he wanted a relationship that was meaningful relativistically.

The reason this distance is meaningful is the fact that, according to special relativity, all inertial observers will agree on it. We have already called it an invariant. If two relatively moving inertial observers look at the same pair of events and calculate the distance between those events using formula (4), they will each obtain the same numerical result. Each observer will measure a different value for dx, dy, dz, and dt, but when these values are squared and combined as in (4), their answers will have the same numerical value. This is what makes Minkowski's expression important. It is not that it is a distance in the usual sense of the term; it is the closest thing to the usual definition of a distance that is the same for all inertial observers.

As I have already mentioned, the invariance of this Minkowskian metric, as it is called, is a result of the nature of the Lorentz transformations. They in turn express the fact that the speed of light is the same for all inertial observers. We might say that three-dimensional Euclidean geometry is necessitated by the fact that all observers who are rotated relative to each other must discover the Pythagorean relationship when measuring the hypotenuse of a right triangle and expressing its length in terms of sides constructed in their coordinate systems. Four-dimensional Minkowskian geometry is necessitated by the additional fact that all inertial observers must measure the same speed of light. Thus the nature of each type of geometry is dictated by a fundamental "truth," which must be experimentally verified, Euclidean geometry by a truth about space alone and Minkowskian geometry by a truth about spacetime. Of course, Euclid thought that his truth was universal, but it turned out that it only held in "flat" space. Similarly, Minkowskian geometry will be shown to hold only for "flat" spacetime. From now on, when we talk about flat spacetime, we will mean a spacetime with Minkowskian geometry, which in turn implies the conditions under which special relativity holds.

One of the strange and unexpected characteristics of the Minkowski metric is that the square of the distance between two points in spacetime can be positive, negative, or zero. We can easily accept positive distances, since that is the usual situation for the distance between two points in space; but how can two distinct points in spacetime be separated by a distance whose square is negative or zero? In fact, if the square of the distance were negative, the distance itself would be imaginary, and how could that be?

To answer this question, we have to look more carefully at spacetime distance by studying how is it measured. Since both space and time are involved, it is clear that distance cannot be measured by a ruler alone. Let us first consider the case in which the two points are separated by a positive distance. This simply means that $(cdt)^2$ is greater than $(dx)^2 + (dy)^2 + (dz)^2$. Since the two points represent events, things that are happening at a time and place, the metric is telling us that light will travel farther than the spatial distance between the two events in the actual time that separates them. This is easy to understand. Take two events that are a fairly large spatial distance apart and apply the Minkowski distance formula. Of course, I am cheating a bit by extending the formula from a small distance to a large distance. I can do this only because, in this example, spacetime is flat and the distances are measured along straight coordinate lines. In a curved spacetime, which I will shortly discuss, the Minkowski formula cannot be extended to large distances so easily; the extension from small to large distances requires integration.

Suppose the two events are (1) you leave San Francisco at noon on a plane heading for New York and (2) you arrive in New York six hours later. Clearly, light could travel from San Francisco to New York in much less time than six hours, so the distance in this case is positive. We see then that events separated by a positive spacetime distance are those that could be connected by an observer traveling slower than the speed of light. In other words, the same observer could be present at each event without ever traveling faster than the speed of light. We call the distance between such events "timelike" and say that the events have a timelike separation. As the observer who connects the two events, you will determine the distance to be the elapsed time on your watch, which is why we call the separation "timelike."

If, on the other hand, the two events are (1) you board a plane in San Francisco at noon and (2) someone else gets off a plane in New York at precisely the same time, then the square of the spacetime distance between these events is negative, because there is a zero time difference between the two events, during which light could travel no spatial distance at all. These sorts of events, which no observer traveling at the speed of light or less than the speed of light could possibly connect (be present at), are said to have a "spacelike" separation.

Finally, there is the special case in which the spacetime distance between two events is zero. This can happen only if a beam of light is present at each event. For example, event (1) is I board a plane in San Francisco at exactly 12:00 noon, event (2) is you, stationed in New York, watch me board the plane on a TV screen. Assuming the video picture is transmitted to New York at the speed of light, the time delay between my boarding and your observing me board is precisely the time needed by the image to cross the continent. Thus the spacetime distance between the events is zero. Whenever two events are exactly connected by light, they are said to have a "lightlike" separation.

Proper Time

Any inertial observer can determine the spacetime distance between pairs of events with two separate measurements. The spatial portion is measured with a ruler and the temporal portion with a clock. Actually, the ruler will have already been used to place coordinates along the reference frame, and a large number of properly synchronized clocks will be positioned at convenient locations. If the two events have a timelike separation, an inertial observer can always be found for whom the spatial separation of the events is zero. In the case of the plane trip to New York, the observer for whom the spatial separation is zero is the traveler on the plane. In the frame of reference of that observer, which is the plane itself, there is no change in position of the two events; they both occur at the door of the plane. Of course, to keep the observer truly inertial for the entire trip, we have to assume the plane does not accelerate, which would be an impossibility. We could, however, choose as our two events two positions in the flight when the plane is already moving with a constant speed. For the observer that moves with the events and is present at both of them, the spacetime distance has no spatial portion. Therefore, the four-dimensional spacetime distance is just what the observer's clock reads (multiplied by the speed of light).

The time recorded by a clock carried by an inertial observer who moves between events is called the proper time interval between those events. It is not to be confused with time intervals obtained from separate clocks positioned throughout spacetime as part of the coordinate system laid out by some arbitrary observer. Those clocks are said to read "coordinate time." To avoid confusion, it is traditional to use different symbols for coordinate time and proper time, the Greek letter tau, τ, for proper time, and Latin t for coordinate time. When we get to curved spacetime, keeping the distinction between these two times in mind will become crucial.

In the Newtonian conception of reality, there would be no difference between coordinate and proper time. A clock, be it worn by an observer, positioned on the mantle of a fireplace, or attached to the wall or floor of a room, should read the same time in all cases.

Imagine you set out on an airplane trip. You set your watch to match the reading on some stationary clock in the airport. Then you board the plane and take off. During the trip, your watch records some amount of elapsed time. Finally, the plane lands and you disembark. Entering the terminal, you once again check your watch against a clock on the wall. Assuming that you have crossed no time zones and that all the timepieces involved are accurate, you would fully expect your watch to read exactly what the terminal clock reads. That is precisely what Newton would conclude.

In the Einsteinian universe, this would not be the case. The airport clocks are fixed in the inertial reference frame of Earth and read coordinate time (which would also be the proper time for earthbound observers standing alongside the clocks). The watch on your wrist is recording your elapsed proper time. The difference between the two airport coordinate-time readings will not equal the proper time that has elapsed for you. What will be equal is the spacetime distance between the two events of your entering and leaving the airplane. That equality will be agreed upon by both you and an earthbound observer who we assume is inertial. For you, the spacetime distance is entirely your elapsed proper time. For the earthbound observer, however, that same spacetime distance is composed of the coordinate-time difference and the spatial distance, combined according to formula (4). Consequently, the coordinate time difference, by itself, cannot possibly equal the elapsed proper time. Newton would be dumbfounded!

Curved Spacetime Compared to Flat Spacetime

In the completely flat spacetime of special relativity, distances between both nearby and widely separated events will be the same to all inertial observers. In the curved spacetime of general relativity, however, only the distances between infinitesimally close events must have the same value, a value that is the same for all observers at the positions of those events, regardless of the way they set up coordinates in their own frames of reference. However, the particular mathematical form of the infinitesimal length will change in different coordinate systems. In other words, different observers will look at the same small distance and express it differently according to whatever coordinates they are using. Each observer deduces a particular metric. According to the principle of equivalence, a freely falling observer discovers the simplest metric of all, the metric of flat spacetime.

To illustrate with an analogy, suppose a group of people, all speakers of different languages, observe the same object, let's say a cat. The object itself is the same to each person, but their means of expressing it are quite different. An American says "the cat." A Frenchman says "le chat." A Spaniard says "el gato." A German says "die Katze." The actual "thing," the cat itself, is analogous to the invariant four-dimensional length. It is the same to all observers; the way it is expressed in their "coordinates," their language in this case, is quite different. A dictionary can be thought of as a kind of metric that connects the invariant thing with its mode of expression. And just as certain languages allow certain concepts to be expressed more easily, so do certain coordinate systems allow length to be expressed more easily.

Why Can Gravity Be Interpreted as Curved Spacetime?

In the preceeding sections, I showed what is meant by flat and curved spaces and filled in some of the mathematical background that is needed to understand general relativity in the deepest sense possible. I've also retraced Einstein's intellectual odyssey, since he didn't have the proper mathematical tools when he started out. But now that you know what a metric is and how coordinates are used, I can forge the final bridge between mathematics and physics by showing how the physical properties of gravity can be thought of as being a result of the mathematical properties of a curved spacetime.

I'm going to attempt to make Einstein's radical ideas plausible, if not completely intelligible, with two analogies. First, imagine I have constructed a large table whose surface can be given any distribution of temperatures by means of heating elements concealed within it. On the surface of this table, while its temperature is constant, I draw a large circle. I proceed to heat the table so that the center of the circle is cold but its perimeter is hot. Having done all this, I ask an observer to measure with a small metal ruler both the radius and circumference of the circle and deduce from the results any geometrical peculiarities. Here, for the sake of the analogy, I will assume that the observer cannot feel the temperature differences and that the metal ruler will very quickly contract and expand according to its temperature. As the observer lays the ruler out from center to circumference, it will proceed to lengthen as it grows progressively warmer. Consequently the observer will measure a somewhat shorter radius than I actually constructed. The ruler will remain at its maximum length on the circumference and will measure a circumference that is considerably smaller than I drew it. As a result, this circle will not satisfy the usual condition that its circumference equals $2\pi r$. In fact, the circumference will be less than $2\pi r$.

How might the observer interpret this result? If not clever enough to figure out my temperature trick, the observer might conclude that the table was warped. In fact, if the table had a somewhat spherical shape to it, as though its surface were draped over a large sphere, the circle would behave in just the way it did. If cleverer or more suspicious, the observer might ask me for a different ruler, maybe a wooden one. Redoing the measurements with a ruler made of a material that does not expand thermally in the same manner as steel would lead to different results, so the observer would ultimately conclude that it was not the circle at fault, but the rulers. Of course, without access to other rulers, the observer would be unable to discard the curved table hypothesis. The point of this little analogy is to show how the behavior of a ruler can trick you into ascribing properties to shapes and surfaces that really do not belong to them.

Now let's try another analogy, using two-dimensional creatures. Suppose these creatures live in a two-dimensional curved space, that is, actually within (not on) the surface of a sphere. We will imagine that one such creature, a mathematician, is attempting to verify experimentally the basic geometrical proposition that the sum of the angles in a triangle is 180°.

Using a well made two-dimensional wooden ruler (which nevertheless must bend to conform to the curvature of the space it is in), the mathematician draws lines connecting three points in the curved space, forming a triangle. With a two-dimensional protractor, the mathematician proceeds to measure and add the angles, and is shocked to find that the sum of the measurements exceeds 180° by a significant amount.

One way the mathematician can interpret this unexpected result is to imagine that some unknown force has bent the ruler in such a way that the triangle was incorrectly drawn (or that temperature variations are changing the ruler's shape in a bizarre manner). Thinking the former possibility to be true, the mathematician chooses a steel ruler, obviously much stiffer than the wooden one, draws the same triangle using the same three points, and is surprised to get the same sum of angles. Obviously the force pays no heed to the atomic structure of the ruler.

At this point the mathematician has two options: either to believe the existence of a force that produces the same effect on all rulers regardless of their structure (a highly dubious hypothesis), or to believe that the nature of space itself is such that triangles do not obey the geometry of ordinary flat surfaces. I suspect the mathematician chooses the latter hypothesis (because it is simpler and more satisfying) and ultimately becomes aware of the surface of the sphere.

These analogies make different points. First, the properties of geometrical objects depend on the results of measurements made with real, physical, material measuring devices. If these devices are somehow affected by the conditions of measurement, the interpretation of the measurements will be similarly affected. Second, if it can be determined that the measuring devices are not behaving normally, the measurements can be corrected and the true state of affairs ascertained (as in the case of the heated table). However, if all measuring devices behave the same way (as in the case of the curved two-dimensional world), then an observer may have no reasonable alternative but to trust the measurements and deduce that the objects or relationships being measured have unexpected properties.

Einstein noted that the same things happen in our world when the acceleration produced by gravity is measured. Suppose a physicist attempts to measure the effects of gravity on some object at Earth's surface. The object itself becomes a measuring device, just like a ruler. Dropping the object and timing its rate of descent, the physicist discovers that it falls with the acceleration g. Another object, be it larger or smaller, wooden or metallic, falls with the same acceleration g. What are the possibilities? Either gravity is a force that produces an acceleration totally independent of the size and structure of the objects on which it acts, or there is some basic condition of four-dimensional spacetime that requires that all objects fall the same way. Einstein chose the latter option because it was simpler and thus more pleasing to his aesthetic sensibility.

Although Einstein did not go about it in quite the way I just suggested, he did use an analogy and an ingenious chain of reasoning. In his 1916 paper "The Foundation of the General Theory of Relativity," Einstein showed that a certain noninertial reference frame would exhibit the same properties that a mathematician would say could result if the structure of spacetime were curved. Einstein's example was a rotating coordinate system, like a large carousel. He considered what would happen if an observer in this system took a large number of very small, rigid rulers of unit length and laid them out along the radius and circumference of a circle constructed around the system's axis of rotation. If the coordinate system and rulers were not rotating, the number of rulers laid along the circumference would be 2π times the number laid out along the radius. The rotation, however, gives each circumferential ruler a velocity along its length, and therefore it will be contracted relative to the radial rulers, which do not have such a lengthwise velocity. Thus more rulers would fit into the circumference than would fit there when the system was not rotating, and therefore the measured length of the circumference would be larger than $2\pi r$. To the observer, this would be an indication that the ordinary relationships of plane geometry were not holding.

This example is very similar to the example I just used of the circle drawn on a heated table. There too the relationship between radius and circumference of a circle was altered by the measurement process. In my example, the thermal expansion of the ruler changed the measurement. In Einstein's example, the motion of the ruler altered its length according to the Lorentz contraction of special relativity. The crucial difference is that temperature will alter different rulers by different amounts, so the clever observer can discover that there is a difficulty with the measuring instrument. In Einstein's example, however, all rulers are affected in exactly the same manner, so no observer is entitled to blame some peculiarity of a particular ruler for the problem.

Einstein proceeded to deduce, in a similar manner, that clocks placed along the circumference of the rotating circle would move slower than a clock located at the center of the circle. In short, Einstein demonstrated that an observer who chose to make measurements of lengths and times from within a rotating coordinate system would arrive at conclusions that would appear quite strange to an observer in a nonrotating system. This is what we mean when we speak of the curvature of spacetime. An intelligent observer is entitled to conclude that spacetime is curved when all rulers, regardless of their material composition, and all clocks, regardless of the nature of their mechanisms, reveal that the relationships between spatial and temporal measurements do not obey the simple laws of flat geometry.

As he had done with special relativity, Einstein convinces us that the properties of spacetime can be determined only by measuring instruments that are physical objects. Spacetime has no properties other than those we infer from our measurements. To Newton, the properties of space and time are mathematical idealizations, like Plato's transcendent ideals. Space is flat or Euclidean, and time is uniform and linear. It is a space and time that a mathematician might construct mentally and then give to physicists to play with. Einstein says, "Give me a clock and ruler and let me check it out!" He discovers the properties of spacetime experimentally. His is a physicist's spacetime. I said before that Newton took the ideal cosmos described by astronomers and the earthly behavior described by physicists and combined them under a single set of laws. Astronomy merged into physics. Einstein now completed Newton's unification. He took the idealized space and time of mathematicians and the material objects of physicists and combined them into a single curved spacetime of physicists.

Having shown that noninertial observers would be convinced they were living in a curved spacetime, Einstein was able to use the principle of equivalence to relate noninertial effects to those of gravity. He reasoned as follows. If noninertial observers live in a curved spacetime, and if observers in a gravitational field are locally equivalent to their noninertial counterparts, then they too must live in a curved spacetime. This is brilliant reasoning. Einstein could never prove directly that gravity was equivalent to a curved spacetime, because he didn't understand gravity that well. He could, however, prove that noninertial motion was equivalent to curved spacetime by applying the results of special relativity, which he did understand well. All that remains is to relate what happens to an observer in a noninertial system to what happens to an observer in a field of gravity, and to do that Einstein used the principle of equivalence.

Eight

Metrics, Geodesics, and the Schwarzschild Solution

In his 1916 paper, Einstein presented the physics community with a radically new mathematical model that would ultimately come to replace Newton's law of universal gravitation. The model provided profound new insights into the nature of gravity and inspired an entirely new way of thinking, not only about gravity, but about forces in general.

In simplest terms, Einstein's model expressed a relationship between matter and energy on the one hand and the curvature of spacetime on the other. Newton's law of universal gravitation,

$$(1) \qquad\qquad F_{1,2} = G\frac{m_1 m_2}{r^2{}_{1,2}},$$

which was the first law that gave mathematical form to the attraction between objects having gravitational mass, declared gravity to be a "force" between two bodies (of masses m_1 and m_2 and separated by $r_{1,2}$) that acted through a distance and yet had no obvious mode of transmission. Einstein's approach expressed a much more fundamental connection between the distribution of matter and energy in the universe and its effects on the curvature of spacetime geometry. It declared gravity to be an effect of that curvature, inasmuch as particles were compelled to move along certain paths (called geodesics) in curved spacetime. As we have seen, whereas Newton's law was originally meant to be understood in terms of action at a distance, Einstein's new theory was a field theory, and the field described by his equation was the geometrical structure of spacetime itself.

The motion imposed on physical objects by gravity was also explained differently by Einstein and Newton. To Newton, objects were acted upon by the force of gravity and responded, as they did to all forces, by accelerating according to his second law, $F = ma$. According to Einstein, objects in a curved spacetime followed a special set of paths called "geodesics"; a force of gravity was neither implied nor required. Thus the particle trajectory obtained by solving Newton's second law became the equation of a geodesic. To Einstein, the motion of a particle was dictated by the geometry of the spacetime it inhabited. The particle had no choice but to follow the geodesic. Thus to determine particle paths, it was not necessary to be concerned with the particles themselves. One simply determined the geometry of spacetime, wrote down the equations of its geodesics, and solved them. As important as they are we will not become involved with geodesics at this point but will rather concentrate on the geometry of spacetime that produces them.

The equation for that geometry that Einstein arrived at relates three very unusual and complicated quantities: the "stress-energy tensor," $T^{\mu\nu}$, which describes the distribution of mass and energy in the universe; the "Ricci tensor," $R^{\mu\nu}$ (named after an earlier mathematician who had studied the properties of the curvature of space), which describes the curvature of spacetime; and the "metric tensor," $g^{\mu\nu}$, which defines the differential properties of spacetime in whatever set of coordinates the equation is written.

Because the way we describe and interpret the nature of our surroundings is heavily dependent on the use of "coordinates," our choice of coordinates has an enormous influence on the utility of our description. The metric tensor reflects both the effects of our coordinate choice and the actual underlying shape of spacetime that those coordinates will be used to label. A simple spacetime described by complicated coordinates can lead to a metric tensor that is quite complicated in appearance, but so can a complex spacetime that is described by simple coordinates. In short, when one is examining a metric tensor, it is a significant mathematical problem to distinguish the effects of matter and energy (producing true curvature) from the effects of the choice of coordinates.

Einstein's equation is written as follows:

$$(2) \qquad R^{\mu\nu} - (1/2)\, R\, g^{\mu\nu} = \kappa\, T^{\mu\nu}$$

The two Greek superscripts μ and v each take on the values 1, 2, 3, and 4, corresponding to the four coordinates of spacetime. Since these tensors have sixteen components, the single equation is really shorthand for sixteen separate equations. The R is the curvature scalar (a zero-rank tensor), derived from sums of components of the Ricci tensor, and the κ is a fudge factor (better known as a "coupling constant") that relates the magnitude of a given mass to the degree of curvature it creates. Its value would ultimately have to be determined by experiment.

It is not my purpose to describe how Einstein arrived at this equation, which is elegant in the simplicity of its form yet extraordinarily complicated in its mathematical substance. It was an intellectual struggle for him and would be far more of a struggle for us. In fact, Einstein didn't actually derive the equation in a series of logical steps; he created it, albeit after much thought, in one fell swoop, using his immense powers of intuition. What is important for us, however, is to simply see what it looks like and attempt to understand the nature of the description of gravity that it provides.

First, we should realize that the general form of the equation is independent of the particular choice of coordinate system. The equation makes no explicit reference to coordinates at all. Of course, the coordinates are there implicitly, but they are buried within the structural form of the three tensors. According to Einstein, a set of coordinates defines the particular frame of reference an observer chooses to be in, and that choice is immaterial to the essence of the underlying reality. The effect of different coordinates will be only to change the ultimate form of the metric tensor, which will then be determined by solving the equation. In other words, for a given distribution of matter (i.e., for a given stress-energy tensor) one particular choice of coordinate system might be more appropriate than another in that it will lead to a simpler version of the metric tensor. The metric tensor bears the burden of the coordinate choice. Nevertheless, regardless of its particular form, the metric tensor will still contain all the information necessary to describe the effects of mass and energy on the motion of objects. It is precisely these effects that we call gravity. In short, what Newton would have called "the force of gravity," in Einstein's terms is only our interpretation of how an object's path in spacetime (the geodesic) is bent by the curvature of spacetime. And that interpretation is itself a product of the coordinates we use, which is to say, the frame of reference we place ourselves in.

The curvature of spacetime is contained within the metric tensor, regardless of its complexity due to particular choices of coordinates. Part of the tensor's complexity is due to the choice of coordinates, and a portion of that part will be what the observer interprets as the noninertial effects (if, for example, the observer happens to be on a carousel or Ferris wheel while attempting to describe the flight of a bird). Another part of the metric tensor will reflect the "real" gravitational effects—that is, those due to the actual mass and energy distribution. That part is ultimately independent of coordinate choice and is encapsulated in yet another tensor, called the Riemann curvature tensor, $R_{\alpha\beta\mu\nu}$. Only when the Riemann tensor is zero can we say that there is no actual curvature present; only then will there be a complete absence of the so-called tidal forces that are characteristic of true gravitational fields. The problem is that a real observer, cast adrift in some coordinate system in the midst of masses and energies, would be hard-pressed to distinguish which effects are coordinate-caused and which are matter/energy-caused. Only the physicist who writes down the equation, knowing the actual distribution of all the matter and energy in the universe and actually choosing the coordinates, has the ability to distinguish these differences.

More about Metrics

General relativity is simultaneously a theory of geometry and a theory of physics: it is a theory of how physics creates geometry and how geometry then affects physics. The early geometers like Euclid studied certain fundamental features of space, particularly the relationships between straight lines and points. Lines and points were idealizations, in the sense that neither a perfect line nor a perfect point could really be constructed. Nevertheless, Euclid believed that the properties of space could ultimately be determined. He believed that there was only one kind of space, the three-dimensional analog of a flat surface. Euclid didn't ask how space got there; he assumed, no doubt, that it was provided by the gods. He simply studied relationships in it. Neither did he question why three-dimensional space had to be flat, even though he knew that two-dimensional surfaces didn't have to be flat.

It is very easy to accept the existence of curved surfaces. Just look at an apple. It is much more difficult to conceptualize a curved space, since, where space is concerned, there seems to be nothing to look at. It was Einstein who realized that "looking" is only one particular way humans perform measurements. We shine light on an object and interpret its properties by studying the reflections with our eyes. Measurements needn't involve the eyes, however. We can also measure the properties of objects with rulers, clocks, and other instruments. Those are the instruments that became Einstein's eyes, with which he "looked" at space and discovered that it need not be flat.

Many centuries after Euclid, less than a hundred years before the time of Einstein, the nineteenth-century geometers Gauss, Bolyai, Lobachefski, and Gauss's student, Bernhard Riemann, had already begun to think about curved spaces and their possible properties. This did not mean that such spaces need actually exist physically, but only that they could be conceptualized and studied mathematically. Gauss studied the geometry of two-dimensional surfaces. He was revolutionary in restricting his attention to small regions of such surfaces and in realizing that a surface can be studied "intrinsically," apart from the higher-dimensional space in which it exists. Gauss showed that the structure of a surface is determined by its metric, the set of relationships between the actual measured distance separating nearby points and the formula for that distance given in terms of coordinates drawn on the surface.

In contrast to Euclid and Gauss, Riemann realized that flatness was only an attribute of certain spaces and that there was a veritable infinity of curved spaces, just as there was an infinite number of curved surfaces. Riemann studied the effects curvature would have on the properties of points and lines that Euclid and Gauss had determined for flat spaces and curved surfaces. Following his teacher Gauss, Riemann concluded that in order to study curved spaces, one should restrict one's attention to very small regions, where relationships can be established that would be extremely difficult to deduce by looking at the spaces on a larger scale. This led to the branch of mathematics called "differential geometry," the geometry of small portions of spaces and surfaces and the relationships between very small lines (called differential line segments) and between points that are very close together (separated by differential distances). Perhaps Riemann's greatest achievement was the study of geodesics. A geodesic is the straightest line that can be constructed in a curved space, the curved-space analog of the straight line in flat space. But even Riemann, for all his brilliance, never ventured to ask, What makes spaces curve? He was satisfied to determine the geometry of given curved spaces.

The Schwarzschild Metric

It is a quirk of history that the first meaningful, rigorous solution of Einstein's equation was not obtained by Einstein himself but by another physicist, Karl Schwarzschild, in 1916, shortly after Einstein's paper was published and just before Schwarzschild himself died. Schwarzschild chose to study a universe devoid of all matter except for a single, spherical mass so small that it could be considered a mathematical point. As it happened, Schwarzschild was able to determine the metric for this universe without actually solving Einstein's equation in its most general form. The fact that only a single pointlike mass was present made the situation so simple that the metric could be obtained by the use of the symmetries implied by the spherical shape of the mass. This is very fortunate, because as simple as Einstein's equation is in appearance, mathematically it is very difficult to solve. It is a tribute to Schwarzschild's cleverness that he was able to "solve" it without solving it!

Schwarzschild chose to describe his single-mass universe in terms of a set of spatial coordinates that mathematicians usually call "spherical polar coordinates" when they are applied to the more usual task of labeling points in ordinary space. We have already used these coordinates in chapter 7 to describe the surface of a sphere. As we shall see, however, the coordinates don't behave quite as one would expect when they are used to label a spacetime containing a mass. Indeed, most of the interesting features of the Schwarzschild metric are due to the way it forces us to reinterpret the "simple" coordinates. To fully appreciate Schwarzschild's metric and its interpretation, we will first look at the way he might have proceeded to determine the meaning of coordinates if there were no mass present at all. This would be the case of an empty spacetime, in which the results of special relativity would hold true and four-dimensional distances would be measured by the Minkowskian metric obtained earlier.

If Schwarzschild were using spherical polar coordinates to label empty spacetime, a spacetime devoid of either matter or energy, he would proceed as follows. He would first select an origin from which to extend a radial coordinate, r, to label positions. He would then select angular coordinates, θ and φ, to measure latitude and longitude around the origin. Finally, t would represent time at all positions in the coordinate system. Think of Schwarzschild attaching symbolic clocks to the very fabric of spacetime. Do not worry about how these coordinates are assigned values; at this point they are simply labels or mental images. In particular, do not think of the clocks as being physical clocks; they can be any mechanism displaying a sequence of numbers.

With the task of constructing the coordinate system completed, Schwarzschild would have to move about within this system and, using a real clock and ruler that he carried with him, make direct measurements of the distance and time between various pairs of events. The physical nature of space and spacetime must be determined by real measuring devices; only they can determine its properties. Coordinates are arbitrary; measurements are real. The metric connects them. Only by making measurements could Schwarzschild verify that the spacetime metric is the simple Minkowskian metric expressed in three-dimensional polar coordinates. We know in advance that Schwarzschild must discover that this is the correct metric, because we already know that special relativity holds in empty spacetime, and the Minkowskian metric describes that circumstance. The metric written in (3) is equation (1) of the previous chapter rewritten in spherical polar coordinates.

(3) $$(ds)^2 = (cdt)^2 - (dr)^2 - (rd\theta)^2 - (r\sin\theta d\phi)^2.$$

In (3), ds is the only expression that means anything empirically. It is the four-dimensional spacetime distance between neighboring events that Schwarzschild would have to measure with a clock or ruler; all the other quantities are merely symbols, like interstate highway numbers or the numbers on their exits, using coordinate values arbitrarily imposed on spacetime to label the events. It would ultimately be up to Schwarzschild to give these symbols meaning, by proving with his ruler measurements, for example, that the distances between points measured radially or in some angular direction are consistent with all of the intuitive notions implied by such a set of spherical polar coordinates.

You could think of this entire process as being analogous to the way in which language might have originally developed. The sounds we utter have no intrinsic meaning; we make them only because they are easy for our throats and vocal cords to produce. After we make a sound, we point to an object, and only then does the sound acquire its meaning. In a very real sense, Schwarzschild would do the same thing. His "sounds" were a set of four coordinates. His metric would give them meaning by "pointing" to actual things, in this case distances and durations between events.

The coordinates chosen by Schwarzschild are "global coordinates"; they are defined in all of spacetime. The major purpose of these coordinates is to give us a framework in which to operate. The metric relationship in (3) is "local"; it is defined in a small region around a point in spacetime. Real physical measurements are always made locally, since our bodies and our instruments are quite small. We can match our local measurements to the values of global coordinates in small regions of spacetime, and in that way relate what are initially mere labels to actual measured quantities. We had better be prepared to discover some strange and exciting discrepancies when we try to reconcile global assumptions with local measurements.

Once again, I would make a linguistic analogy. English is universally accepted today as a global language. Nevertheless, if we were to take a document written in English and present it to people in various other countries, we would discover that they give meaning to that document by translating it into their own languages. We can make interesting deductions about their native languages by comparing the way they interpret the same English document. Once again, this illustrates how fundamental differences can arise between global and local descriptions, but a global description can still serve to connect local ones. I will show how this also happens in general relativity.

Schwarzschild would have to give meaning to the coordinates that globally label empty space by making physical measurements. We, however, will perform these measurements in our imaginations only, allowing the metric itself to provide us with the answers we believe Schwarzschild would get with his ruler and clock.

Do the Coordinates Really Mean What They Say?

Imagine a sphere, floating in space, centered on the origin of coordinates. Using the radial coordinate symbol r as a label, we will assign this sphere a "radius" R. Every point on the surface of the sphere is labeled with the same value, $r = R$. According to what we know about flat three-dimensional space, there should be a simple relationship between the radius of a sphere and the circumference, C, of its equator: $C = 2\pi R$. We have to show that Schwarzschild would obtain this result if he actually measured the circumference of our imagined sphere. Fortunately we don't really have to measure the circumference, because the metric tells us the result we would obtain if we did.

In order to make correct deductions from the metric, we have to show how it can be used to simulate a ruler measurement. Remember that the symbol ds represents the spacetime distance between two events. A ruler measurement can be expressed as this distance if we call an observation of the position of each of its ends an event. With this way of looking at things, when a ruler one meter long is placed with zero velocity, the simultaneous observation of both its ends produces two events that define the ruler's proper length. The events are a meter apart in space and zero distance apart in time, and the ruler used to produce these events is said to have a proper length of 1 meter. The use of a 1-meter-long ruler to make a measurement in this way therefore corresponds to setting $(ds)^2 = -1$. I will disregard units for now. The negative sign indicates that, in Minkowski's terminology, ds is a spacelike distance. Since we will ultimately need to make spatial measurements much smaller than one meter, let's give Schwarzschild a ruler of differential length dl, which he will use in every measurement that follows.

We can now proceed to measure the circumference of our sphere's equator, which is to say that we count the number of times the ruler must be laid down, end to end with zero velocity as we simultaneously observe both its ends, so as to completely encircle the sphere. To use the metric in (3) as a surrogate for the measurement process, we fix the time coordinate so that $dt = 0$, keep r fixed at its value of R so that $dr = 0$, and choose $\phi = 90°$ to place us on the equator. This simplifies the metric to

(4) $(ds)^2 = -(R d\phi)^2$.

We can simulate Schwarzschild's ruler-placing trip around the equator by letting ϕ increase from $0°$ to $360°$, which is 0 to 2π radians. (Note: when one is using polar coordinates, angles must always be expressed in radians rather than degrees, a radian being an angle unit defined so that $360°$ is 2π radians.)

Each time Schwarzschild lays down his ruler, observing both its ends simultaneously, it measures a spacelike distance $ds^2 = -dl^2$ and subtends an infinitesimal angular width $d\phi$. As he moves completely around the equator, therefore, the ruler will cover the entire 2π radians of that angle. Referring to our metric in (4), we obtain the total circumference by integrating

(5)
$$\int dl = \int R d\theta = 2\pi R$$

for the complete trip around the sphere. Note that the 2π in front of the R comes from the fact that the complete 360° trip is also a trip of 2π radians.

Because each placement of the ruler defines a length, dl, whose square is $ds^2 = -dl^2$, we can conclude from (5) that Schwarzschild would have to lay the ruler down exactly $2\pi R$ times to circumscribe the equator. That means that the relation between the radius and circumference of our imaginary sphere is exactly what we expect from Euclidean geometry. Thus we are almost at the point of being able to say that the global symbol r does indeed have the properties of a normal radius in spherical coordinates.

Let us verify one more fact. This time we will imagine a second sphere, concentric with the first but radially larger than it by the small amount dr. This sphere will be labeled by the coordinate value $R + dr$. We must now determine how many ruler lengths will be needed to stretch radially between the surfaces of the two spheres. In terms of the metric, we keep dt, $d\phi$, and $d\theta$ all zero and find the relationship between ds and dr. From (3) we now have

(6)
$$ds^2 = -dl^2 = -dr^2.$$

This simply says that dl, the ruler length, and dr, the coordinate difference between the spheres, are exactly the same. Now we can state that the radial coordinate r is exactly what we suspected it to be, the usual spherical radius.

To complete the interpretation of coordinates, we must finally allow Schwarzschild to provide a meaning for t. To do this, he compares the time between two ticks (the events) as read by his personal clock, which we have called "proper time" or $d\tau$, with the time that would be registered by the mechanism reading coordinate time. Although we shall shortly see that this is not a simple comparison when a mass is present, in the empty-universe case it is. We can see from the metric, after setting all coordinate differences equal to zero (since Schwarzschild is standing at one place) that

(7)
$$(ds)^2 = (cd\tau)^2 = (cdt)^2,$$

or $d\tau = dt$.

In short, the proper time, $d\tau$, that elapses between ticks of Schwarzschild's real clock is the same as the coordinate time between ticks, dt, that is produced by the coordinate clock mechanism. Therefore, Schwarzschild can identify coordinate time, the global label, with his personal clock readings made locally. Until this comparison is made, we can't say anything about the nature of coordinate time other than that it uniquely labels events that are also separated in proper time. Now we can state that coordinate time can actually be defined by real clocks, fixed in space. We can synchronize such clocks, moreover, just as we did when discussing special relativity.

Armed with the technique for giving meaning to coordinates by using the metric, we can do exactly the same thing for the far more interesting case of a single mass in otherwise empty space. In this case, however, the Minkowskian metric is not applicable, so we need the correct one, which is the one Schwarzschild actually obtained as a solution to Einstein's equation.

The Metric of Spacetime Containing a Single Mass

The result Schwarzschild obtained has come to be called the "Schwarzschild metric" or the "Schwarzschild solution":

$$(8) \qquad (ds)^2 = (1 - 2m/r)(cdt)^2 - (1 - 2m/r)^{-1}(dr)^2 - (rd\theta)^2 - (r\sin\theta d\phi)^2.$$

Admittedly this is not a very simple expression. It is certainly more complicated than the Minkowskian metric that we have just been examining. Moreover, we have no way of knowing if the coordinate labels r, t, θ, and ϕ have the same meaning as they do in flat spacetime, because Schwarzschild made only the most basic assumptions about their properties. We shall see that they do not have the meaning we expect and that this spacetime cannot be interpreted as being Euclidean. It is our first example of a spacetime with curvature.

Let me first point out that the constant m in the Schwarzschild metric is the product of the actual mass M of the object and the constant G (Newton's universal gravitational constant) divided by the square of the speed of light. In other words, $m = GM/c^2$. We shall analyze the role m plays in specific cases.

To gain insight into the physical and geometrical nature of Schwarzschild spacetime, we will do exactly what we did previously: attempt to give meaning to the global coordinates by making local measurements. Let's first look at the spatial portion of the metric by itself. We will allow Schwarzschild to use his ruler to determine the meaning of the coordinate r. Using the same approach as before, we imagine a sphere of radius R centered on the origin, which will now contain the mass. We watch as Schwarzschild walks around the sphere, repeatedly dropping his unit ruler and measuring the circumference of the sphere's equator by counting how many times he drops the ruler. We will obtain the result for ourselves by using the metric. We can set $dt = 0$ for the spatial measurement, which gives us a much simpler looking metric:

$$(9) \qquad\qquad (ds)^2 = -(1 - 2m/r)^{-1}(dr)^2 - (rd\theta)^2 - (r\sin\theta d\phi)^2.$$

As before, Schwarzschild is making all measurements with a ruler whose proper length, defined as the length obtained by a simultaneous observation of both its ends by an inertial observer at rest relative to it, is exactly dl. We cannot assume that the ruler can be held fixed in spacetime in the presence of gravitational forces and still be dl units long (even though "dl units" still appears on the ruler). However, according to the principle of equivalence, if the ruler were short enough and were allowed to fall freely in a gravitational field, it would be unaffected by that field and behave as though it were in an inertial reference frame. Furthermore, since $(ds)^2$ between any pair of points (events) is independent of the coordinate choice labeling those points, it must retain its numerical value whether expressed in Schwarzschild's coordinates or the coordinates of a freely falling frame of reference. Mathematically, we mean that although the ruler's spacetime length is defined by the value of the interval between its ends when it is at rest and inertial, $ds^2 = -dl^2$, we can also equate that numerical value to the expression for its spacetime length, even in the presence of a mass, as long as it is momentarily at rest while freely falling. The falling ruler can then be used to measure the distance between points labeled by a fixed coordinate system. In what follows, Schwarzschild will be comparing proper length measurements to coordinate difference measurements.

The Measuring Process

Having Schwarzschild walk around the equator of the sphere is equivalent to setting $\theta = \pi/2$ radians (or 90°) in (8), because that is where the equator is located. Also, since he remains on the equator, we set $d\theta = 0$. Since the coordinate radius of the sphere is fixed at R, we must also set $dr = 0$. Finally, we have already set $dt = 0$, because Schwarzschild observes both ends of his ruler simultaneously, meaning that each measurement takes literally no time at all. Imposing all these mathematical restrictions on the metric and noting that $\sin \pi/2 = 1$, the metric in (9) simplifies to:

(10a) $$ds^2 = -dl^2 = -R^2 d\phi^2,$$

or

(10b) $$dl = Rd\phi.$$

Were Schwarzschild to make a complete round-trip around the origin so that his angle of longitude, ϕ, increased by exactly 2π radians, the mathematically equivalent process would be an integration, $\int dl = \int Rd\theta = 2\pi R$, to obtain the length of the circumference. If he therefore places the ruler down (allowing it to fall freely) a sufficient number of times to encircle the origin, he will measure a cumulative length of $2\pi R$. This is precisely what we'd expect along the circumference of a circle whose radius is R. It tells Schwarzschild that the geometry of this space is normal in the sense that the radius and circumference of a circle maintain their usual Euclidean relationship. In this regard the result is not different from what we already verified previously with the Minkowskian metric when no mass was there.

Schwarzschild's next task is to consider whether the geometry of space is Euclidean in the radial direction as well as in the tangential direction around the equator. He can do this by considering a second sphere, concentric with the first but displaced from it by a small coordinate amount dr. Now he lays his ruler down in a radial direction from one sphere to the other, comparing their separation as measured by the ruler with the coordinate difference indicated by the two labels on the spheres, R and $R + dr$. From the metric we see

$$(11) \qquad (ds)^2 = -(dr)^2/(1 - 2m/r) = -(dl)^2,$$

which relates any radially measured spacetime distance dl to its corresponding coordinate difference dr, giving

$$(12) \qquad dr = dl\sqrt{1 - \frac{2m}{r}}.$$

Equation (12) tells us that the ruler, which is dl units long in spacetime distance, occupies a coordinate distance dr of $dl\sqrt{(1 - 2m/r)}$ units. Here, for the first time, we see a discrepancy between a physical measurement and our intuitive sense of the nature of the set of coordinates we are using to provide us with a description of spacetime. Under ordinary circumstances, we would expect that the infinitesimal distance dl measured by the ruler would simply equal the coordinate difference dr marked off numerically along the radial coordinate axis. Every measured step of one meter should pass one meter's worth of coordinate markers, but this doesn't happen. According to the metric, if Schwarzschild were to walk toward smaller and smaller values of r (but no farther than $r = 2m$, about which I will have more to say), placing the ruler down as he does so, the expression for dr in (12) would become an increasingly smaller fraction of the length (dl) of the ruler, so it would take more and more ruler placements to mark off fixed distances in r. Therefore, he either concludes that the ruler is getting progressively shorter the closer he gets to the mass, or else that space is somehow being stretched out before him like taffy, because the distance between successive coordinate markers is getting greater.

The only way I can give you a sense of this strange behavior is to have you imagine that you are in a car, heading toward a city straight ahead of you. You are watching both the car's odometer (which registers dl) and the mileage markers alongside the road (which register dr). To your amazement, as each additional mile is registered on the odometer, you pass smaller and smaller fractions of a mile in terms of the mileage markers. You begin to question either the odometer, the markers, or your sanity.

If Schwarzschild were a student of curved spaces, he could rationally conclude that he was in such a space. An example of how a curved space could produce the same strange results is illustrated in figure 8.1. The funnel-shaped space is a two-dimensional space that you can think of as inhabited by two-

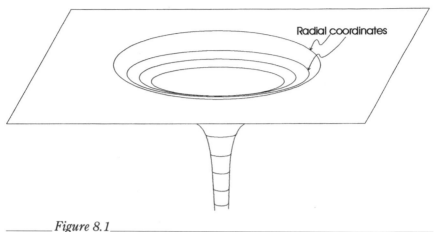

_____ *Figure 8.1*_____
A Funnel-shaped Space

dimensional creatures. The spheres of constant radii in Schwarzschild's space become circles in this space. A creature placing a ruler between successive circles must place the ruler in the surface of the funnel. To that creature, the circles are uniformly spaced in radial distance, but they are not uniformly spaced along the surface of the funnel. Yet it is the surface of the funnel within which the creature exists and within which the creature is making measurements. The radial coordinate is, in effect, an illusion. It is only the measurements made within the fabric of the funnel that are real. When the creature lays down the ruler in this surface, it must be laid down an increasing number of times to cover the separation between circles that are equally separated in radial coordinates. As can be seen from the figure, the farther inward the creature goes, the more vertical is the ruler placement and the more ruler placements become necessary to cover the distance between circles. Of course, the creature cannot "see" the funnel-shaped space. The creature sees only the ruler and the circles.

What Is Time?

There is an equally strange behavior of the time portion of this spacetime, and understanding it will require more of an effort on our part. Whereas previously we equipped Professor Schwarzschild with a differential ruler, we will now give him a differential proper-time clock, that is, a physical clock that ticks once every $d\tau$ seconds when at rest in an inertial reference frame. Such a clock can also be defined as one that produces a timelike spacetime interval between ticks whose length is $cd\tau$ meters (since $ds = cd\tau$).

Schwarzschild's first task will be to move about spacetime, comparing the proper-time interval between ticks of his clock when it is inertial to its value expressed in terms of t, the global unit of coordinate time (i.e., the time read by the fixed-coordinate clocks used to label the coordinate system of spacetime).

To evaluate the relationship between his clock and ones mounted along the roadside, Schwarzschild must pick some position in space and compare the elapsed time between ticks of his clock with the corresponding elapsed time indicated by the coordinate clock at that position. As before, we will deduce the result of Schwarzschild's measurement by studying the metric.

For ease of use in the vacuum of outer space, in which sound does not carry, we assume that both Schwarzschild's clock and the coordinate clock are constructed to emit a flash of light with each tick. Let t_1 and t_2 be the times of two successive flashes as produced by the coordinate clock at (r, θ, ϕ). We assume the flashes are close enough together so that we can say $t_2 - t_1 = dt$. Schwarzschild now lets go of whatever he has been holding on to and allows himself to fall freely past the fixed clock, arranging things so that he passes it with a zero velocity, $v = 0$. By falling freely, he brings the principle of equivalence into play and eliminates all effects of gravity in his immediate vicinity. Thus his clock is inertial and is ticking at its manufactured rate, namely $d\tau$ between ticks. It is also important that Schwarzschild have an instantaneous zero velocity relative to the fixed clock. If he did not, the time between its successive ticks, dt, when recorded by his clock would be lengthened by the time dilation of special relativity.

The two successive flashes emitted by the fixed clock are two events whose infinitesimal spacetime separation is the same whether expressed in the inertial (Minkowskian) coordinates of Schwarzschild's freely falling frame (i.e., read off his personal clock) or in the global coordinates of spacetime. It is the basic rule of general relativity that an infinitesimal spacetime interval is the same in all reference frames. The coordinates describing the interval may vary, but the length they express is invariant.

In Schwarzschild's freely falling system, the two flashes of the fixed clock have no spatial separation (since, instantaneously, the clock is at rest relative to Schwarzschild), so their spacetime separation is just c times the elapsed proper time, $d\tau$, on his clock:

(13) $$ds^2 = c^2 d\tau^2.$$

When described in the global coordinates of spacetime, the same separation between ticks must be expressed in terms of the Schwarzschild metric:

(14) $$(ds)^2 = (1 - 2m/r)(cdt)^2 - (1 - 2m/r)^{-1}(dr)^2 - (rd\theta)^2(r\sin\theta d\phi)^2.$$

The spacetime separation between the two ticks of the clock still has no spatial component (because the clock is fixed in spacetime), so (14) simplifies to

(15)
$$ds^2 = c^2\left(1 - \frac{2m}{r}\right)dt^2.$$

Since the separation, ds, in (13) and (15) must be the same, we can equate them, take the square root of both sides, divide one by the other, and obtain

(16)
$$\frac{dt}{d\tau} = \left(1 - \frac{2m}{r}\right)^{-1/2}.$$

Since the ratio in (16) is not equal to 1, the time between successive ticks, dt, produced by the coordinate clock is therefore not the same as the time between those same ticks as indicated by the elapsed proper time on the freely falling clock held by Schwarzschild, $d\tau$. In fact, the ratio depends on the position of the coordinate clock. Schwarzschild's real interest, however, is not the comparison of his real clock and the coordinate clock at one point of spacetime, but the comparison of the rates of two real clocks at two different points of spacetime. Let us proceed to investigate that comparison.

Suppose we have two identical clocks, each in free-fall at different radial positions of spacetime. One clock is the one we just investigated, in free-fall at r. The other clock falls freely at a more distant radial value, r'. The relationship between coordinate time and the time registered by the more distant zero-velocity, freely falling clock at r' can be obtained by inserting the new value of radial position:

(17)
$$\frac{dt}{d\tau} = \left(1 - \frac{2m}{r'}\right)^{-1/2}.$$

Notice that if this second clock were fixed at an infinite radial distance from the origin, we would obtain the result $dt = d\tau$. It is only at infinite distances from the central mass that coordinate time and physical-clock time are identical. An observer at infinity would be justified in identifying coordinate time there with the proper time recorded on a personal clock. For this reason, coordinate time in the Schwarzschild universe is sometimes referred to as the proper time of an infinitely distant observer. This also means that such an observer can note the coordinate time of events at any other points in spacetime and establish the time differences between them simply by subtracting the corresponding times recorded on that observer's own clock.

We can now deduce the results of a simple but important experiment, a comparison of the elapsed time between successive ticks of a real clock freely falling (with zero instantaneous velocity) at r with the elapsed time between ticks of a more distant but identical freely falling clock at r'. We will use coordinate time to establish the comparison because it has the convenient property of being global. It is the physical proper time that is important; the coordinate time is used only as an intermediate step to establish the connection. To make this comparison, we assume that Schwarzschild positions himself at the more distant r' clock and looks back at the r clock. The clock at r flashes at coordinate time t_1, and the light from that flash arrives at Schwarzschild's position some coordinate time Δt later, at coordinate time $(t_1 + \Delta t)$. That is when Schwarzschild sees the flash. The time delay, Δt, is the amount of coordinate time it takes the light to travel a radial-coordinate distance $r' - r$. It is very important to remember that because coordinate time is global, it is the same everywhere. Something that happens at time t at one place also happens at that same time t relative to every other place in spacetime. This is what allows you to say that the coordinate clock at the place that Schwarzschild sees the flash is actually reading $(t_1 + \Delta t)$. The clock at r then flashes a second time, t_2, and that light reaches Schwarzschild at $(t_2 + \Delta t)$. We assume that the travel time is the same in each case, since no changes have occurred in the structure of spacetime during Δt.

The elapsed coordinate time between flashes as observed by Schwarzschild at r', $dt(r')$, is

(18) $$dt(r') = (t_2 + \Delta t) - (t_1 + \Delta t) = t_2 - t_1 = dt(r),$$

which is exactly the elapsed coordinate time between the flashes at their emission points. However, Schwarzschild will be reading proper time on a real clock, so we must deduce the elapsed proper time he will observe at r', $d\tau(r')$, given that the coordinate-time difference there is $dt(r)$. We can obtain this result from equation (17):

(19) $$d\tau(r') = dt(r)\sqrt{1 - \frac{2m}{r'}}.$$

At r, the point of emission of the two light signals, the relation between coordinate and proper time durations is

(20) $$d\tau(r) = \sqrt{1 - \frac{2m}{r}}\, dt(r).$$

The ratio of $d\tau(r')$ to $d\tau(r)$ is

(21) $$\frac{d\tau(r')}{d\tau(r)} = \frac{\sqrt{1 - \frac{2m}{r'}}}{\sqrt{1 - \frac{2m}{r}}}.$$

Since r' is greater than r, $d\tau(r')$ is greater than $d\tau(r)$. This means the more distant clock shows that more time has elapsed between the two flashes than the clock at the position of their emission. Thus the more distant clock is running faster. In other words, Schwarzschild will see the clock at r, closer to the mass, running slow relative to his clock at r'. This phenomenon is called gravitational time dilation.

If, instead of a flashing clock, we use two identical atoms that radiate light at a given frequency, the atom at r will radiate at a lower frequency than the atom at r'. The frequency of an atomic transition, f, is inversely proportional to the time between successive vibrations; thus

(22)
$$\frac{f(r)}{f(r')} = \frac{\sqrt{1 - \dfrac{2m}{r'}}}{\sqrt{1 - \dfrac{2m}{r}}}.$$

This phenomenon is now called the gravitational red shift.

The "Schwarzschild Radius"

The strangest behavior of the Schwarzschild metric occurs in the region of spacetime where $r = 2m$. At this value of the radial coordinate (which defines a complete sphere), the metric implies that a physical clock stops moving relative to another clock that is radially farther away, and a ruler can be moved radially inward without reducing the value of r by a single centimeter! This value of r, an impenetrable barrier impervious to both clocks and rulers, has come to be called the Schwarzschild radius.

Schwarzschild himself was puzzled and disturbed by the behavior of spacetime at $r = 2m$. He tried to content himself with the following reasoning. First of all, the value $r = 2m$ turns out to be extremely small. Using Earth as an example, we can calculate m from the values 6×10^{24} kilograms for the mass, 6.67×10^{-11} Nm^2kg^{-2} for G, and 3×10^8 ms^{-1} for c. These values give us 4.44×10^{-3} meters for m and twice that for $2m$. This is a bit less than one centimeter. Most important, it is seemingly buried within Earth. Moreover, you can't reach this radius by digging a hole because the metric is no longer valid within Earth's interior. There Einstein's equation would be entirely different, because the body of a planet is a region filled with matter rather than devoid of matter. Beyond Earth, the Schwarzschild metric would be correct, but it can be evaluated only at r values far greater than $2m$, at which positions nothing very unusual happens. Within Earth, the metric is completely invalid.

Moreover, Earth cannot be shaved or peeled and made smaller so that the mysterious value $r = 2m$ is exposed. If mass is in any way removed, $2m$ also gets smaller, so the strange radius seems always to be hidden deep within the mass that produced it. The only way the Schwarzschild radius could ever be observed would be if the mass producing it were actually packed within a sphere less than $2m$ in radius. Earth would have to be compressed to the size of a marble, with all of its mass intact. Schwarzschild could therefore be content with the knowledge that $r = 2m$ was an unattainable radius; consequently, the strange predictions of his metric seemed to be mathematical rather than physical. But were they?

Schwarzschild's comfortable conclusions were shattered by the results of a paper published in the *Physical Review* in 1939 by the eminent physicist J. Robert Oppenheimer and a colleague, H. Snyder, entitled "On Continued Gravitational Contraction." Since the paper was published at the beginning of World War II, it languished in the journal where it appeared, without gaining much attention. Oppenheimer himself was soon involved in the Manhattan Project, so it is doubtful he did much to publicize the paper. Eventually, however, it came to light, and then it had a remarkable effect.

What Oppenheimer proved was that a star, in the natural course of its development, could exhaust its nuclear fuel and collapse in on itself while still maintaining virtually all of its original mass. The physical radius of the star would shrink until it reached the Schwarzschild radius of $r = 2m$, at which point that special value would be "exposed," together with all of its strange attributes. If such a process actually occurred, it raised two very troubling questions. First, how would the collapse appear to the eyes of a distant observer? It would seem that the surface of the collapsing star would never actually reach the critical radius, whose advent would take longer and longer over the last few units of coordinate distance. Second, how would the collapse appear from the standpoint of an observer on the surface of the star itself? Would this observer also never reach the $r = 2m$ value, or in fact cross over into some nether region of spacetime?

The answers to these two questions will serve to illustrate how strange an object a collapsing star is, and even more, how strange are the properties of spacetime in the vicinity of such a star. Before looking at those answers, however, it is worth taking a few moments to look at the collapse process itself, because the physics would be interesting even if spacetime behaved in a more rational manner.

The Death and Ultimate Collapse of a Star

In Oppenheimer's theory, a star's death and collapse occur in three stages and has three possible outcomes. A normal star, like our Sun, is in equilibrium under two opposing forces: its own gravitational self-attraction, which tries to pull it inward, and the internal pressure of radiation and hot gases, pushing it out. The outward pressure is due to the heat and radiation produced by nuclear reactions within the core of the star; the gravitational attraction is due to the mass of the star and the spacetime curvature it produces. The first stage of collapse occurs when the nuclear fuel in the star's core becomes exhausted.

The core begins to cool and collapse. The collapse is a result of the imbalance between gravitational force and thermal pressure as cooling occurs. As the core collapses, there is a conversion of gravitational energy to motional energy of the matter in the core, and therefore the collapse releases heat. This heat flows outward from the solid core into the mantle of gases that surrounds the core and ignites the hydrogen in the mantle. Now a strange transformation occurs. The burning mantle begins to expand, even as the core continues its collapse. The dying star thus swells enormously into what is known as a red giant. Our own Sun will ultimately become such a red giant, in the process engulfing most of the solar system. The collapsing core, meanwhile, reaches a size so small that a new force emerges, countering the effects of gravity and halting the collapse. This new force is due to the nature of the matter that makes up the core, nuclei that are left over after all nuclear reactions have ceased occurring (mostly nuclei of iron, the most stable of all known nuclei) and all the electrons that surrounded the original nuclei when they were still in their atomic state. The electrons are now much too energetic to remain attached to a nucleus, so they form a gas, which surrounds the remaining iron nuclei.

Electrons belong to a class of particles called fermions, which are extraordinarily antisocial; it is virtually impossible to get two of them close to one another. To begin with, they exert a repulsive electrical force on each other, which strengthens with decreasing separation. They also operate under a quantum mechanical dictum, called the Pauli exclusion principle, that literally forbids any two of them to occupy the same "state." That means that no two electrons can simultaneously have the same set of physical attributes, such as position, momentum, and direction of intrinsic rotation (called "spin"). Since there are so many electrons in the collapsing core, most of the attributes have been spoken for, and the only thing keeping electrons in distinct states is the fact that they are physically separated from each other.

A gas of electrons, in which all available attributes have been taken, is called a "degenerate gas." As the core attempts to collapse, however, the electrons are being forced to come closer and closer together, a process that, because of their "degeneracy," is strictly forbidden. This "forbidden" behavior shows up as a colossal pressure within the gas that forces the electrons apart in an effort to keep them in distinct states. "Degeneracy pressure," as it is called, is much stronger than the electrical repulsion between electrons (which is not nearly strong enough to combat gravity), and it is also stronger than the gravitational force trying to collapse the core. In short, the collapse is halted. The bloated mantle ultimately stops burning, cools down, and is pulled back to the now stable core. This last process isn't always benign, and sometimes the mantle can bounce around a bit and even blow off some of its gases in explosive events. Whether or not this occurs, however, the red giant shrinks considerably and ends its life as a small, dim star called a white dwarf.

As it turns out, however, a white dwarf does not always form. If the original star is more massive than 1.4 times the mass of the Sun, the pressure of the degenerate electron gas will not be sufficient to hold back the pull of gravity. This 1.4 factor is called the Chandrasekhar limit, in honor of the brilliant astrophysicist who calculated it in 1935 (even before the work of Oppenheimer). If the limit is exceeded, the core continues to collapse despite the wishes of the electrons. It was this possibility that Oppenheimer investigated in his paper of 1939.

If the collapsing core of the dying star is between 1.4 and 3 times the mass of the Sun, a different fate lies in store for it. Cores of this size are typically left behind by stars that are initially much larger than the Sun, maybe thirty times as large. These stars too go through the red-giant phase of their death, but the more massive burning mantle of gases also begins to collapse as it cools, and because of its size the collapse leads to a catastrophic explosion called a supernova. The supernova blows away much of the huge cloud of burning gases that surrounded the star and contained what remained of its heat. What is left is the cooler core of the star, a cinder from the original nuclear burning that maintained the star during its normal life. This core contains much of the mass of the star, although little of its heat. Once again, gravity becomes the dominant force, and now, because the Chandrasekhar limit is exceeded, the degenerate-electron pressure cannot withstand its pull.

As the electrons get squeezed together by the force of gravity, seemingly beyond the ability to resist posited by the Pauli exclusion principle, they pursue another course of evasive action. They are forced into the protons still residing in the iron nuclei, interacting with them in a process that produces neutrons. This process is signified in the following way:

$$p + e^- \rightarrow n + v.$$

In words, a proton (p) plus a negative electron (e^-) turn into ($\rightarrow$) a neutron (n) and a neutrino (v). The neutrino, I might add, doesn't stick around but quickly leaves the star, heading for interstellar space.

Once this reaction has taken place and most of the electrons and protons have been replaced by neutrons, the core continues its collapse. The neutrons, however, have other plans. They too are fermions and must obey the Pauli exclusion principle. Their degeneracy pressure is even stronger than that of the electrons. Once again the collapse is halted, leaving behind a core of pure neutron matter called a neutron star. I must point out that pure neutron matter is so dense that a thimbleful would weigh 100 million tons on Earth.

The condition of a neutron star is not always a terminal state. Oppenheimer, in a slightly earlier *Physical Review* paper, "On Massive Neutron Cores," written with G. M. Volkoff, showed that if the neutron star has a mass greater than about 0.7 times the mass of the Sun, it will be unstable and continue to collapse. The so-called Oppenheimer-Volkoff limit represents the maximum mass that a neutron star can have if it is to remain a neutron star. Since the Oppenheimer-Volkoff limit is less than the Chandrasekhar limit, a star passing from a white dwarf to a neutron star would have to lose some of its mass to remain stable.

What happens to the star that becomes a neutron star but is too massive to remain that way? It continues to collapse, but now there are no new pressures that can arise to counter gravity and halt the collapsing process. The star has become so small that its gravitational force (more correctly, the curvature of spacetime) exceeds the strength of anything else nature could throw in its way. The collapse therefore proceeds until the matter in the core is squeezed into nothingness! The star presumably shrinks to a size that is less than its Schwarzschild radius and becomes, in the jargon of modern astrophysics, a black hole.

This is the way in which a physical object could achieve the size of its own Schwarzschild radius, a possibility that Schwarzschild himself must have dreaded, since the radius marked a region of spacetime whose behavior seemed too preposterous to believe. We are forced to consider it, however, now that Oppenheimer has given us a physical mechanism by which it could happen.

If you will recall our discussion of the Schwarzschild metric, the strange behavior of spacetime in the vicinity of $r = 2m$ was the fact that clocks slowed down to the point where time itself appeared to stop and rulers seemed to indicate that it would be impossible to reach that position in space. We made these strange deductions by considering Schwarzschild's own choice of a global coordinate system and comparing it to local measurements. What would happen if a different choice of coordinates were made? All we have really discovered, after all, is that in terms of those coordinates spacetime behaves very strangely. It needn't behave strangely in another set of coordinates. The coordinates, after all, embody Schwarzschild's vision of the proper language to use in discussing spacetime, a perfectly fine language when spacetime was empty and flat. Conceivably spacetime's paradoxical behavior is really the fault of Schwarzschild's coordinates and not of spacetime itself. Certainly spacetime is not completely innocent. It is, after all, not Euclidean; it is curved. And we should expect certain of its features to be strange to us. But should they be quite as strange as Schwarzschild's metric predicts?

The Ride of Your Life: Falling into a Black Hole

One rather nice feature of the Schwarzschild metric is that at large distances from the mass, the coordinate time label, t, and the proper time of an observer's clock, τ, are identical. This can be seen directly in equation (16), by letting r become large so that the square root is essentially equal to 1. I used the analogy before of coordinate labels and interstate route numbers in our highway system. At great distances from the mass, it is as though the highways that differ in route number by one unit were actually one mile apart. You might imagine one clock tick near the mass leaving along route 1, followed by another tick along route 2. The ticks remain on their routes until they finally arrive at the edge of the universe, where a seated observer finds that they are one second apart. Therefore, a distant observer can simply read the time on a wristwatch, which is both proper time and coordinate time, to obtain the coordinate time in that distant part of the universe. This fact will become important for us.

Let's imagine that such a distant and cautious observer would like to learn about the properties of spacetime in the vicinity of a black hole. Being much more cautious than curious, the observer asks an adventuresome friend to assume all the risk by sitting on the surface of a star that has begun to collapse. We will have to use our imaginations and knowledge of the Schwarzschild metric to describe what would occur. The observer asks the friend to send back simple messages by light signals, perhaps just wave a flashlight back and forth to indicate that all is well.

Assuming that the star is fairly large when it begins its collapse, the fearless adventurer is initially quite far from its Schwarzschild radius, and therefore the beam from the flashlight appears to have the same frequency as light emitted by a comparable flashlight in the hands of the cautious and distant observer. Therefore, as the intrepid traveler waves the flashlight signaling that all is well, the distant observer sees exactly that.

Now the collapse begins in earnest. The star begins falling inward to its own mysterious destiny. As the radius, $2m$, is approached, the frequency of the light emitted from the friend's flashlight begins to be less than the light from the observer's flashlight. To the distant observer, the light seems to grow dim, because more and more of the frequencies that it emits are red-shifted, falling below the threshold that the eye is sensitive to. In addition, the rate at which the friend waves the flashlight also seems to slow down, as does the rate at which the friend's heart beats and wristwatch ticks.

As judged by the distant observer, the friend never reaches the Schwarzschild radius but moves ever so slowly, grows continuously dimmer, and finally freezes completely in the coordinate time of the observer, remaining a distant black image for the rest of time. So, in fact, does the entire star! The black hole, in the eyes of the distant observer, is a dim and totally unmoving object, floating forever, or at least until the end of the universe, in the void of spacetime. The Schwarzschild radius is commonly referred to as the "event horizon" because it marks the position of the last events that a distant observer will ever see.

How does our intrepid traveler describe the same voyage? One would think the description would be substantially similar. So much for naive intuition. To describe the voyage from the standpoint of the traveler, we need to use the traveler's own proper time as the descriptive coordinate, not the coordinate time of the distant observer. If you were calling home to describe an automobile trip you were taking, you would tell how many miles you had traveled, not how many route numbers you had passed. Here is where general relativity and the curvature of spacetime rear their ugly heads.

As it turns out, the traveler not only crosses the event horizon (passes the Schwarzschild radius) but does so in a finite amount of time as measured by the accompanying watch. This amazing result was independently discovered by the physicists Finkelstein and Kruskal, who demonstrated that with a different choice of coordinates than Schwarzschild made, one could show that the horizon was indeed crossed and that the traveler would continue on an inexorable (no turning back is possible), painful, and exceedingly brief trip to $r = 0$, the origin of the coordinate system. From a mathematical point of view, they discovered that the Schwarzschild radius is not a true "singularity" of nature or a barrier to further travel. It is merely an artifact of the coordinate system that Schwarzschild chose to use. The point $r = 0$, on the other hand, is a legitimate singularity; it is a point beyond which nothing can go.

There are some interesting things about this scenario that should be emphasized. First of all, although the event horizon is not a barrier to physical travel, it is a barrier to communication. Across the horizon, no messages from the external universe can reach the traveler. The second fascinating thing is that an infinite amount of time has already elapsed in the external universe by the time the event horizon is crossed. Whatever the fates had in store for our cautious observer has already occurred. The third thing is that the crossing of the event horizon, while easily done, is a far from painless process. As the traveler approaches the horizon, the spatial curvature becomes so extreme that the traveler's body is stretched beyond its physical limits. Described in terms of Newtonian gravity, we could say that the force at the traveler's feet (assuming a feet-first fall) is much greater than the force at the head. Thus the traveler is not simply being pulled inward as a whole but is being stretched in the process, with feet accelerating more rapidly than head. We have already called this type of differential pull of gravity a tidal force because it is responsible for the solar and lunar tides here on Earth. Although the curvature caused by the Sun at Earth's surface is far less than that caused by the nearness to an event horizon, the size of Earth is so much greater than that of a person that the curvature effect is noticeable as a pulling of the waters in the oceans.

Black Holes and the Meaning of Existence

Do black holes really exist? This is a trick question because it involves an ambiguity in the word "exist." As we have just seen, a collapsing star produces the phenomenon we call a black hole by progressively exposing its Schwarzschild metric until the infamous event horizon at $r = 2m$ is almost visible, yet to an observer in spacetime the exposure is never completed. The star gets closer and closer, but never reaches that mystical radius. So to the curious but cautious observer, the black hole—which is to say, the $r = 2m$ position, is never actually reached. The black hole, therefore, does not exist.

The fearless traveler falling inward with the collapsing star has no such experience. Not only does the black hole exist, it destroys the star and the traveler in a wink of an eye and goes out of existence. So for that person, the black hole has existed but no longer exists.

How can the black hole both not yet exist and no longer exist? The reason is that the black hole is not a "thing"; it is a condition of spacetime. In our mesoscopic reality, things are objects that have existence. Existence is an absolute quality; it is not dependent on the point of view of the observer. In general relativity, however, different frames of reference can differ so markedly in their spacetime properties that the very concept of "To be, or not to be" goes beyond even Hamlet's conception of the import of that question.

Nine

The Very Many

Statistical Reality

In the second half of the nineteenth century, educated men and women had developed a certainty that science and its methodology would play an increasingly important and beneficial role in shaping the future of humanity. Science was built on a multilayered foundation of certainty. There was certainty about the value of knowledge and therefore about the search for it; there was certainty about the virtually unlimited ability of human mental faculties to find that knowledge and use it wisely; and there was certainty about the predictability of physical phenomena. These certainties were not the result of a single flash of insight, but were the legacy of two hundred years of assimilating and appreciating the successes of Newtonian physics. Some felt that science had begun to replace religion in the hearts and minds of men and women as the road to ultimate truth and understanding. Knowledge was now to be obtained by observation, measurement, and analysis rather than religious revelation. Closely associated with this view was the feeling of an inevitable, continuing, and intrinsically beneficial "progress," which was to be a major by-product of a world built on the foundation of science.

It is hardly surprising that this pervasive spirit should manifest itself in positivism, the all-embracing philosophy of certainty that we encountered as an influence on Einstein. (The name was coined earlier in the nineteenth century by the French philosopher Auguste Comte.) Although the spirit of positivism pervaded all aspects of intellectual life, the core of its strength resided within the natural sciences. Two of the leading exponents of positivist science were the Austrian physicist Ernst Mach and the English physicist William Thomson,

Lord Kelvin. The brilliance of these two scientists and the respect accorded them gave immeasurable support to the positivist principles they espoused.

As we have seen, Mach was particularly known for his desire to eliminate from scientific theory all concepts not directly related to sensory information. A science that was of the senses, by the senses, and for the senses should not require the use of ideas that were removed from the senses. The concept of an atom, for example, should be unnecessary in a legitimate scientific theory because the atom could not be directly sensed. Any theory that professed to explain a phenomenon perceived by the senses by positing the presence and behavior of insensible objects like atoms and molecules was unworthy of the adjective "scientific." In Mach's own words,

> The ultimate unintelligibilities on which science is founded must be facts, or, if they are hypotheses, must be capable of becoming facts. If the hypotheses are so chosen that their subject can never appeal to the senses and therefore also can never be tested, as is the case with the mechanical molecular theory, the investigator has done more than science, whose aim is facts, requires of him—and this work of supererogation is an evil. (E. Mach, *Conservation of Energy* [Chicago: Open Court Publishing Co., 1911], 57)

Mach was also distressed by what he sensed as a growing reliance by scientists on the use of mathematics, not as a tool for problem solving, but as a method of conceptualizing the physical world. Mach, of course, was an excellent mathematician, but he used mathematics to solve problems that arose in physics. Some physicists, however, went beyond that, using mathematics more broadly to suggest possible ways of interpreting the world around them. In other words, they used mathematics as a source of "models" of the world rather than a deductive tool. Mach distrusted these mathematical and ultimately mental models as much as he distrusted hypothetical concepts such as ether and atoms.

It is doubtful that anything disturbed Mach and his followers more than developments in the branch of physics known as thermodynamics. It was in this area that two of his most despised scientific heresies were being committed, the application of hypotheses that went beyond the realm of the senses and the use of mathematics to provide a model of reality. Both of these heresies could be laid at the feet of another physicist whom we have already met, Ludwig Boltzmann. Needless to say, Boltzmann and Mach were not on the closest of terms, personally or intellectually.

It is not my intention in this book to dwell on personalities or to provide a detailed history of the scientific developments that occurred during this period. What I will do, however, is explain from a modern scientific perspective what Boltzmann was trying to accomplish and the degree to which he succeeded. It is largely as a result of his success that we have been forced to confront one aspect of our ignorance of physical reality.

Heat and temperature are two of the most important concepts in thermodynamics. It was long recognized as an empirical fact that under certain well-defined circumstances, objects became hotter or colder when they were placed in contact with other objects that were themselves already respectively hotter or colder than the original objects. It was accepted as undeniable physical law that, under these circumstances, placing an object in contact with something hotter would raise its temperature, while placing it in contact with something colder would lower its temperature.

A simple explanation was offered for this phenomenon. The change in temperature was accomplished by the transfer of a certain quantity of heat from the hotter object to the colder one. This heat took the form of an invisible but material fluid, called "caloric," whose total amount could be neither increased nor decreased and which flowed from the hotter to the colder of the two objects. The caloric was the material embodiment of heat; once inside the colder object, it produced an increase in temperature. The mechanism through which the caloric acted was not understood, so scientists were content to confine themselves to making experimental determinations of the amount of caloric required to produce a given temperature rise in specific materials. For example, the amount of caloric needed to raise the temperature of one gram of water one degree Celsius, was called one "calorie," which then became the basic unit of research, not to mention the future bane of dieters everywhere.

Although they were a minority, there were scientists who had misgivings about the caloric theory and sought an explanation for temperature changes through other mechanisms, such as the motion of internal constituents of matter. In 1798 the American scientist Benjamin Thompson (later awarded the title "Count Rumford") claimed to have proved experimentally that heat was not an indestructible material fluid but could be endlessly created by friction. Following experiments that involved the drilling of bores in cannon barrels, he wrote that heat "cannot possibly be a material substance," and that nothing "is capable of being excited and communicated in the manner the heat was excited and communicated in these experiments except it be motion." One year later, the English chemist Sir Humphrey Davy demonstrated that rubbing two blocks of ice together caused them to melt, proving to him that heat was caused by friction. Davy wrote that it was the motion of the blocks of ice that produced the heat and that heat itself must therefore also be a form of motion. He considered the possibility that it was "a peculiar motion, probably a vibration of the corpuscles of bodies."

Neither Thompson's nor Davy's work was sufficiently convincing to destroy the caloric theory, but it subsequently led to a more definitive series of experiments by the English scientist James Prescott Joule, which did. In 1840, Joule showed that heat could be produced by the expenditure of mechanical work, which was already known to be a form of energy. He went on to establish the conversion factor between calories and mechanical work, showing that the work required to raise a one-pound weight 772 feet, if dissipated as friction, will raise the temperature of one pound of water by one degree Fahrenheit.

Ernst Mach would have approved of the experimental results of Thompson, Davy, and Joule, since they eliminated the need for a hypothetical (and now superfluous) fluid, caloric, and instead explained thermal phenomena purely in terms of the transfer of energy between objects. One "extra" concept, caloric, is therefore eliminated from the structure of physics. That is clearly an economy of thought.

On the other hand, Mach would probably have been less pleased by the conclusions of Thompson and Davy that the transferred energy produced its temperature increase by stimulating the motion of "corpuscles" of the matter into which it entered. Mach in fact preferred to think simply that heat was energy and that energy increased temperature. He did not see the necessity of the intermediate hypothesis that the energy was first converted to motion, be it the motion of corpuscles, molecules, or atoms.

The First Law of Thermodynamics

The efforts of Thompson, Davy, and Joule ultimately led to a simple but profound statement concerning the effects energy can have on material objects. Called "the first law of thermodynamics," the statement is usually cast in the following words: The net heat energy entering a material object (the heat that goes in minus what goes out) will numerically equal the mechanical work done by the object plus the change in the internal energy of the object. This law states both that heat and mechanical work are equivalent forms of energy and that the total amount of energy involved in a physical process does not change (is conserved). Heat should be thought of as the particular form of energy that moves between objects at different temperatures, mechanical work is the orderly application of forces to or by objects (such as the friction force involved in rubbing two blocks of ice together), and internal energy is the generic term for energy that, to Davy, resides in the corpuscular motion within an object, but to Mach would simply be something basic within matter. Note that the concept of temperature has not yet been defined. For the purposes of the following discussion, it is sufficient to think of temperature as the reading of a thermometer.

The Second Law of Thermodynamics

If energy can produce a change in temperature, is the opposite also true: can a temperature difference produce energy? Certainly a temperature difference can produce the flow of energy called heat. But this fact is not all that interesting. There is another form of energy, work, that is much more interesting and important than heat. Work is a form of energy that is associated with useful activity; it is pushing something or pulling something. The precise physical definition of work is that it is a product of a force and the distance the force moves in its direction of application. When Humphrey Davy rubbed two blocks of ice together and melted them, he was converting the work of his own muscles (the force they exerted on the blocks of ice, multiplied by the distance they moved the blocks) to heat, and the heat thereupon entered the ice and produced a temperature change.

To what degree can a given amount of heat produce this useful form of energy called work? This, it turns out, is quite a complex question, and the answer to it is so important that it is called the second law of thermodynamics. The second law of thermodynamics puts the answer in the form of a statement about what would be impossible for any sort of device that operates in a cycle (i.e., a machine or an engine) to do: It is impossible for a device operating in a cycle to extract heat from a hot object and convert it to an equivalent amount of work." What makes this law fascinating is that, intuitively, there seems to be no reason for it. If heat and work are both forms of energy, why can't they be completely transformed, one to the other, in either direction? Boltzmann would look to the molecular nature of matter for the answer to this question.

Prior to Boltzmann's efforts, the second law of thermodynamics was simply accepted as a reflection of the way the world behaves. Even though heat and work are both forms of energy, heat was considered by Lord Kelvin (in 1851) to be a "degraded" or "lower-quality" form of energy as compared to work. Thus while the first law defined what happened to the amount of energy involved in a physical process (it remained unchanged), the second law was interpreted as being a statement about what happens to the quality of energy in a process. It becomes degraded.

In 1867, the physicist Rudolf Clausius created a new quantity, which he called "entropy," that could be used to provide a measure of energy degradation during specific natural processes. According to the definition proposed by Clausius, the change in the system's entropy was the quantity $\Delta Q/T$, where ΔQ is the amount of heat entering or leaving a system during a physical process and T is the temperature of the system while the process occurs. For this definition to make sense, you have to assume that ΔQ is infinitesimally small, so that T remains constant over the entire system during the process.

Using the second law of thermodynamics, Clausius showed mathematically that the change in entropy of a system plus its surroundings (which, taken together, is everything that exists) can never be negative. In short, when things happen, the entropy of the entire universe increases or remains the same. Thus Clausius's definition of entropy change could be interpreted as being equivalent to Lord Kelvin's somewhat looser notion of energy degradation. An increase of entropy during a physical process means that the energy residing in the universe is degraded so that less of it is available to do work. It is quite impressive that the original form of the second law, which simply states that no cyclical device can convert heat completely to work, is equivalent to a statement about the quality of energy in the entire universe. We have Clausius to thank for demonstrating this amazing result.

If you asked a physicist of Mach's era why it is that work can be completely converted to heat but heat cannot be completely converted to work, the answer would be that only in the former case does the entropy of the universe increase. However, in a sense this is not really an explanation; it is merely a restatement of the original fact in mathematical terms. The expression for entropy change, $\Delta Q/T$, is a ratio that Clausius discovered will increase for all the "right" processes, the ones that actually occur. While this is a glorious discovery in itself and a tribute to Clausius's genius, it does not provide enormous insight into the reason heat will not become work. Rather, it restates the fact with mathematical precision and elegance. If a teacher were asked why one student performed better on tests than another, that teacher might respond that the superior student's IQ was greater. But that doesn't explain as much as one might suppose. The term "IQ" is defined by a student's performance on certain tests, so any explanation of superior test performance given in terms of a higher IQ is like saying, "That student performs better on tests because that student is a better performer on tests." It sounds ridiculous when said that way, but the term "high IQ" is simply a way of saying just that. Similarly, the "higher entropy" of the universe is the result of a permissible natural process, while "lower entropy" is not, which is a bit like saying that permissible processes are those that occur!

Boltzmann's Microscopic Description

Boltzmann was dissatisfied with the second law in the form stated by Clausius, because he believed that a still deeper insight was possible. For Boltzmann, that deeper insight required a molecular description of matter rather than a description couched only in macroscopic or mesoscopic terms such as "temperature" and "heat," which conveyed no deeper level of meaning. A molecular description, on the other hand, would ultimately depend on the masses and velocities of the molecules involved, and would thereby penetrate to the most fundamental level of understanding.

The problem that Boltzmann faced was enormous. The number of molecules in even the smallest system is almost beyond the mind's capacity to conceptualize. Yet any truly microscopic description would have to deal with the behavior of all those molecules. On the other hand, the quantities of interest to the scientist, the variables that were actually measured in the laboratory, were quantities like heat and temperature that seemed almost disconnected from the molecular details of behavior. Boltzmann would somehow have to connect microscopic behavior, which was barely conceivable to mesoscopic and macroscopic behavior, which was actually measurable.

The Problem of Time Reversibility

Apart from the philosophical problem of connecting one reality to another (which we shall deal with in great detail), there was a second difficulty of physical and mathematical significance. Presumably, any description of the motion of molecules would have to make use of Newton's laws of motion. These laws possess a mathematical symmetry called "time reversibility," which we have already touched on briefly in our discussion of symmetry in chapter 4: an equation (like Newton's second law) has this fascinating symmetry if its mathematical form does not change when the variable denoting time, t, is replaced by its negative, $-t$. Now what exactly does this mean? It sounds as if we were contemplating making time run backward, which seems a somewhat unreasonable, if not ridiculous, thing to attempt. To a mathematician, changing the sign of t in an equation does not necessarily mean "making time run backward." A mathematician thinks of t as just a variable that allows the motion of a physical object to be followed. In that sense, it is not different from the numbering of pages in a book. When the time variable increases, we say that the object is moving forward in time; when it decreases, we say the object is moving backward in time. With this interpretation, moving backward in time could be visualized by playing a motion picture or videotape of the object's motion in reverse,

If we think in terms of reversing a videotape, then changing the sign of t in Newton's equation produces a description of whatever motion would actually be seen on the reversed tape. Furthermore, because Newton's equation doesn't change its form when the sign of t is changed, any motion that you could see on a videotape running in reverse would be a motion that you could also see on a videotape running forward. All time-reversed motions are also perfectly reasonable forward-time motions, because they are solutions of the same equation. We assume, of course, that motion behaves according to Newton's laws.

The problem facing Boltzmann in trying to create a molecular description of mesoscopic phenomena was the fact that Newton's laws are time reversible, whereas mesoscopic phenomena are not. The irreversibility of real-world processes is precisely the subject matter of the second law of thermodynamics. The fact that work can be converted entirely to heat but heat cannot be converted entirely to work is the central irreversibility of the mesoscopic reality we live in. Any observer transported to a world in which time ran backward would see miraculous processes in which heat was converted completely to work. These would be the same processes that, in the ordinary world, are perfectly commonplace examples of work being transformed to heat. Thus a pair of ice cubes rubbed vigorously together and melting would, in a time-reversed world, be an amazing ice-cube vibrator, powered by the spontaneous freezing of a layer of water between them.

Beyond such obvious work/heat conversions, the second law of thermodynamics applies to all natural processes, including life itself. Perhaps the strongest way of stating that the mesoscopic world's thermal processes are irreversible is to use Clausius's entropy. The very fact that there is a quantity (the entropy of the universe) that must increase during natural processes is an immediate indication that the universe contains a fundamental irreversible element. The "direction" of entropy change, the fact that the entropy of the whole world can never decrease, gives a direction to time. In the properly time oriented world, entropy increases or at most stays the same. A motion picture running backward can always be detected through changes in entropy in the scenes. In this sense, a clock can be thought of as an entropy meter. If there is a clock in a scene, its movement is the clearest indication of the way the film or video is running.

Although we usually don't think much about it, clocks really perform two very different functions. The regularity of their ticking mechanism—their "rhythm," so to speak—allows a unit of time duration to be defined. This is the function of a clock that Einstein was concerned with. We can define the interval between two successive ticks as one second, for example, and measure elapsed time or duration by simply counting the number of ticks. This is the use of the clock that we ordinarily make; it is the only reason that we use clocks at all. However, the clock also performs a second operation: it gives us a direction to time. Since our world contains no regions in which time moves in any other direction than past to future, we never have the need to use the directional property of a clock and are not in the least bit conscious of it.

Although the directional aspect is virtually useless, we get it free of charge when we buy a clock. The clock mechanism is designed so that it operates only in one sense. Thus the usual analog clock rotates only in the "clockwise" sense, while a digital clock produces sequences of increasing numbers. You could produce a clock that runs counterclockwise, but that would still be its forward direction. The point is that clocks can't change their direction; as long as a clock moves in its normal direction, we say that time is going forward. Of course, it's really the clock mechanism going "forward," but that's the only way it can go, so it reinforces our intuitive notion that it's also the only way time can go. Any mechanism designed so that it can go only one way must operate according to physical principles that would decrease entropy if the mechanism spontaneously reversed itself. (We are not allowed to reverse the mechanism, because then we become part of it.) In a mechanical clock, the one-way mechanism is called an "escapement"; in an electronic clock, the mechanism is far more complicated, but it is guaranteed to produce sequences of increasing numbers. In this way, a clock measures time direction, which is to say, it defines the increase of entropy.

Now what does all this have to do with molecular motions? Can a horde of molecules, moving strictly according to Newton's laws, be trained to perform as a clock? At first glance, the answer seems to be a resounding "No." Time reversibility of molecular motion forbids it. But Boltzmann looked beyond this intuitive conclusion. His idea was the following. If we have a very large collection of moving molecules such as a bottle filled with a gas, their motion would consist of straight-line paths constantly interrupted by collisions. After each collision, the two colliding molecules would move off on new straight-line paths until another collision occurred. This behavior is perfectly consistent with Newton's laws. Therefore, if at any time during their motion the molecules could be stopped and turned around, they would proceed to retrace their paths, collisions and all, until they returned to some previous configuration. Suppose you made a motion picture of the behavior of such a system of molecules for a very small period of time, like one second. If you ran the film backward or forward, the motions projected on the screen would look virtually identical. Nothing occurring in the reversed film would seem impossible.

Suppose, however, that the bottle had initially been empty and the gas now filling it had been introduced through a valve in a small puff. It then spread out quickly to fill the entire bottle. If that were the case, running the motion picture backward for a long enough time would eventually show the molecules reforming into the small puff. This sequence of events would be extremely unusual, and you would probably conclude that the film was running backward. The ability to discern a difference between the two directions of the film requires both that the initial condition of the gas be somewhat unusual (e.g., a small puff of gas) and that we observe the film for a long enough time to see this initial condition being approached.

This kind of reasoning enabled Boltzmann to arrive at an entirely new conception of the behavior of a gas. He concluded that time-reversed behavior is not impossible, but it is highly improbable. If all the molecules in the gas were to have their motions reversed at some instant of time, they should indeed retrace their steps to the point at which they were introduced into the bottle, but this would involve a perfect set of motions that it is unreasonable to believe would occur in the real world. The film, possessing a perfect memory, simply displays the reverse of what occurred. But in a real situation, any slight disruption of even a single molecule would prevent the complete reversal of the motions. The reason we do not see time-reversed behavior, therefore, is not that it is forbidden by any law at the microscopic level of reality, but rather that it represents a sequence of motions never observed at the mesoscopic level of reality. The second law of thermodynamics, stating what can never occur in mesoscopic reality, simply reflects our interpretation of behavior that is highly unlikely to occur in microscopic reality.

Boltzmann tried to put his ideas into sound mathematical form so that he could convince other scientists of their validity. His first step was to express the motion of the gas molecules with a single equation that took into account both their straight-line motion and the effects of collisions. This was no easy task, so he made some reasonable approximations, assuming that collisions randomized the paths of molecules. In other words, a molecule would be likely to go off in any direction after a collision. With this supposition, Boltzmann was able to derive a mathematical function that, miracle of miracles, would only increase with time! In short, he actually derived something in the microscopic world that behaved like Clausius's mesoscopic entropy function. The critical difference was that, whereas entropy was created simply to have the proper behavior in terms of mesoscopic variables, Boltzmann's function was derived from basic principles and "reasonable" assumptions and couched in microscopic terms.

Unfortunately, Boltzmann's work met with severe criticism. "How can you deduce irreversibility from motion that is reversible?" his critics asked. The answer, in fact, was rather simple. By assuming that collisions randomize the directions of molecular motion, Boltzmann was putting into his equation exactly what he wanted it to give him. He was not deducing irreversibility, but creating it. Let's examine this in greater detail.

When molecules move "forward" in time, we do not care what they do when they collide. They just keep bumping into each other and moving off in some fashion. We have no particular expectations of where they will ultimately wind up, so we see their motion as being random and disorganized. When the same molecules are reversed and asked to move backward to some original state, however, their ability to do so requires that their velocities continually fulfill very precise relations. A pair of time-reversed molecules that collide at time t must emerge from that collision in such a way that each of them engages in the correct previous collision. In the jargon of physics, time-reversed molecules have motions that are "highly correlated," meaning that each molecule in a way knows exactly what the others are doing. By assuming that collisions randomize motion, however, Boltzmann was destroying all possible correlations. The molecules would lose all sense of where they had to be next for the proper sequence of collisions to occur. It is not surprising that Boltzmann obtained a function that increased in time. There would be no way for his equation to describe a situation in which a cloud of gas molecules moved toward an initial puff. It could easily describe a puff of gas that moved so as to fill the bottle, but it would never allow a filled bottle to regress to an initial puff.

Boltzmann's critics made it abundantly clear to him that he had gone wrong. Nevertheless, Boltzmann realized that his equation did in fact appear to describe irreversible reality, even though it produced this description incorrectly from his premises. He hadn't derived irreversibility from Newton's laws, as he had hoped to do, but had derived it from Newton's laws plus the addition of a randomizing effect caused by collisions. The question then became whether nature is irreversible because such a randomizing effect actually exists separate from Newton's laws. Had Boltzmann arrived at a fundamental truth about microscopic reality that correctly explained irreversibility, or had he simply made an incorrect assumption that somehow produced a function increasing in time, like entropy? Boltzmann was put in the position of having to defend an equation that he believed correctly described the mesoscopic world but of lacking physical justification for its derivation. He thereupon set about to produce that justification by reinterpreting the equation.

The Statistical View

Boltzmann's reinterpretation introduced a revolutionary new way of looking at things. It is here that he upset the positivistic thinking of Mach and Kelvin by adding a mathematical insult to the injury he had already caused by his use of a molecular hypothesis. His equation would not describe the behavior of any real individual puff of gas, but would rather describe the behavior of a "typical" puff. In short, the equation became the description of a mathematical concept rather than of a physical object. Boltzmann would agree that molecules whose velocities had miraculously been turned around would proceed to move in a way that reduced their entropy. His equation would not be able to describe that occurrence. But clearly this was not "typical" gas behavior in any sense of the word. A bottle of gas with all its molecular velocities perfectly reversed would be a miraculous entity. Certainly it could not be achieved in real life by the activities of any physicist, however careful. And even if it could be momentarily produced, the slightest deviation at any moment would destroy its subsequent behavior. Therefore such a state of motion could not be considered typical of real (natural and spontaneous) behavior. In order to explain what Boltzmann did, I will use the modern terminology of statistical physics, which we will need to learn for the subsequent chapters on chaos.

Microstates and Macrostates

The microstate of a system, like the bottle of gas we have been discussing, is the complete microscopic description of that system. It requires a specification of the position and velocity of every single molecule in the gas. Taking as an analogy a deck of 52 playing cards, a microstate of that deck would be the specification of the position of every card in it. A microstate is described in terms of microscopic variables, which label the features of the system that are apparent only when it is observed at its smallest scale. The position and velocity of a molecule are such microscopic variables. The position and identification of a playing card are the deck's microscopic variables.

A macrostate, on the other hand, is the description of a system with macroscopic variables. These variables represent the features of the system that we can "sense," either with our own sensory organs or with the proper technological measuring devices. For describing a bottle of gas, the macroscopic variables might include the volume of the portion of the bottle that has gas in it, the temperature of the gas, the pressure of the gas, and the total energy of the gas. These variables each label a characteristic of the gas as a whole and do not refer to its molecular constituents. The macroscopic variables appropriate to a deck of cards might include the number of red cards in the top half of the deck.

The relationship between a microstate and a macrostate is this: every microstate corresponds to some particular macrostate, but a given macrostate virtually always corresponds to many microstates. The number of microstates that are consistent with a macrostate is called the "multiplicity" of the macrostate. Think for a moment of a deck of cards in a macrostate in which all of the red cards are in the top half of the deck and all of the black cards are in the bottom half. This is exactly one macrostate. How many microstates are consistent with this macrostate? The number is unimaginably huge: 26 "factorial" squared, which is written $(26!)^2$ and means $(26 \times 25 \times 24 \times \ldots \times 1)^2$. Every arrangement of the deck that keeps the top half red and the bottom half black is a microstate consistent with that macrostate.

Now we can define a microstate of the gas in the bottle. Such a microstate has to be defined at one instant of time, since the motion of the molecules causes it to change constantly. The microstate will be determined by specifying the exact position together with the exact velocity of each molecule. This amounts to six variables per molecule, three for its position and three for its velocity. If we would interchange only two of the molecules, say molecules number 1 and number 2,054, we would have a different microstate. The two microstates would correspond to the same macrostate, however, as it is highly unlikely that the gross features of the gas would be affected if two molecules changed places.

If we keep all the molecules but one fixed in space and shift that one ever so slightly, we create a new microstate. Similarly, if we keep the velocities of all the molecules but one fixed and change the velocity of that one ever so slightly, we create a new microstate. Since molecules can move anywhere within the bottle and have virtually any velocity, we see what an immense number of microstates can be created. In fact, the only way to conceive of such numbers is to use a bit of mathematical trickery.

Suppose we have only one molecule in the bottle, and at some particular instant it has a given velocity. Each position in the bottle where that molecule could be would constitute a different microstate. If the volume of the bottle is V, we will say that the number of different microstates is also V. This makes sense, because doubling the size of the bottle would double the number of microstates and that is what going from V to $2V$ would do. If there are two molecules in the bottle, then each can be anywhere, and the number of microstates is the product of the number of places the first molecule can be and the number the second can be, which is V^2.

Just as the volume of the bottle indicates the number of positions a molecule can have, we can also imagine a "volume" for its velocity. After all, a volume simply defines spatial limits to where a molecule can be, so we can equally well have velocity limits to how fast a molecule can move. In practice, macrostates usually refer only to some portion of the total volume of the container and a set of limits that determine the velocities the molecules can have.

Phase Space

In order to describe the behavior of a gas over time, which is to say the evolution of its microstate, it is helpful to introduce a new kind of "space" that the gas can be thought to inhabit, called "phase space." This is not the ordinary three-dimensional space that you and I and the gas occupy; this space is a purely mathematical construction, designed to simplify the description of gas behavior. Phase space is a space of many coordinates. In fact, there are three spatial coordinates and three velocity coordinates for each molecule. A phase space for a gas with N molecules therefore has $6N$ coordinate axes. It is a veritable porcupine! The virtue of creating such a cumbersome space is that the microstate of the entire gas is represented in it by a single point. If we observe the evolution of the gas in phase space, we see the motion in time of only one point. This is an extraordinary simplification, enabling us to easily conceptualize certain features of a very complex system.

Whereas a microstate is a single point, a macrostate corresponds to a particular region or volume within phase space. This is a very important distinction, so we should consider it carefully. Every volume in phase space, no matter how small, can contain many individual points. Each point, however, is a particular microstate of the gas. If it turns out that all the points in a certain phase-space volume are consistent with the same macroscopic description of the gas (i.e., they are macroscopically the same state), then they are part of a macrostate. If we can determine the region of phase space that contains all the microstates of a particular macrostate, we are justified in labeling that region as a macrostate.

Now we get to the heart of Boltzmann's conception: every macrostate has an entropy. The numerical value of its entropy is the size of the region of phase space that represents it. When the gas is first puffed into the bottle, it is in a microstate that corresponds to a macrostate with low entropy. That simply means that the initial macrostate is a region of phase space containing relatively few microstates. The reason for this is that the gas initially occupies a small three-dimensional volume, and as we have seen, the number of microstates corresponding to a macrostate is proportional to the macrostate's three-dimensional volume.

As the gas expands, its microstate moves into that part of the phase space containing macrostates corresponding to larger three-dimensional volumes. These macrostates have higher entropy precisely because they correspond to larger volumes. Finally, of course, the gas fills the entire bottle and is in the macrostate of maximum volume, which is the volume of the bottle itself. This macrostate also occupies the largest portion of the phase space and therefore has the maximum entropy. The maximum entropy macrostate is commonly called the "equilibrium macrostate," because the gas is in equilibrium, meaning that its perceptible macroscopic features no longer change. Of course, from the microscopic point of view it is difficult to even conceive of equilibrium, because the gas is in a constant state of motion. But this motion corresponds only to microstate changes, which are imperceptible to mesoscopic creatures like us. We can sense changes only in a macrostate, and when we no longer sense such changes, we are perfectly justified in saying that the system is in equilibrium.

This was Boltzmann's most brilliant insight. Once the microstate of the gas enters the "full-bottle" or equilibrium macrostate, it virtually never leaves. The reason is simple. This macrostate occupies such an enormous region of phase space that the microstate literally can't escape. The gas reaches maximum entropy and stays there. Of course, I should should never say "never." A microstate could conceivably wander out of the equilibrium-macrostate region. But if it does so, it will almost certainly return. Such highly unlikely and brief escapes from equilibrium are called "fluctuations," and they are actually observed and studied by scientists.

Let's repeat this analysis using a deck of cards in place of a bottle of gas. Imagine a brand new deck is removed from its package and shuffled repeatedly. This process can be thought of as corresponding to a small amount of gas being puffed into a bottle and allowed to expand. The shuffles of the deck play the role of collisions between the molecules. When the deck is first removed from its package, like the gas first being puffed into the bottle, it is in a special microstate that belongs to a macrostate of extremely low entropy. I will call this the "new-deck" macrostate. In it, the cards are ordered both by suit and by card value within each suit. Since every new deck of cards is found in this microstate, it constitutes a unique single-microstate macrostate.

Shuffling the deck for the first time produces a new microstate. This microstate belongs to some other macrostate, which is not nearly as well defined as the initial one. Let,s call it a "partially ordered" macrostate, since most of the cards will still be reasonably well arranged by suit and value. Clearly there are many microstates that would correspond to this macrostate. In fact, every possible card configuration that could be achieved with a single shuffle of a new deck belongs to this macrostate. It certainly has many more microstates than the new-deck macrostate and therefore has a higher entropy.

It is not difficult to visualize what will happen as we keep shuffling the deck. After a very few shuffles, the deck arrives at what I will call its "maximally disordered" macrostate. This macrostate contains all microstates that show no vestige of an original order. The suits and values are mixed up; in short, the deck looks like nothing at all. Once it reaches this condition, additional shuffles will not change it. Each new shuffle does indeed produce a new microstate, but it is one that looks just as disordered as the previous one. The maximally disordered macrostate has the highest entropy because there are far more microstates that look like nothing special than there are microstates with some degree of order. Of course, even a maximally disordered macrostate has microstates with some order. That's why people play poker: they hope to be dealt a flush or a full house from a shuffled deck of cards.

Is it possible that repeated shufflings will eventually put the deck back into its original microstate? Of course it is possible, but the likelihood is so miniscule that I would not be unreasonable in saying that it can't happen.

Probability and Phase-Space Volume

Boltzmann's interpretation of the behavior of mesoscopic reality requires that typical systems move in such a way that their microstates eventually enter the macrostate with highest entropy and, once there, become effectively trapped. What about a microstate that is not typical, like a microstate produced by reversing the velocities of every particle of one that is typical? Such a microstate would not behave according to Boltzmann's conception, but it would be so unstable that the slightest disturbance of even one molecule would throw it off course and make it more typical. In short, typical microstates behave according to Boltzmann's equation, and nontypical microstates quickly become typical and are soon trapped in the equilibrium macrostate.

The trapping of a microstate is a result not of physical law, however, but of probability. We make the fundamental assumption that the volume of a region of phase space is proportional to the probability that a physical system will enter that volume and thereby adopt the macroscopic characteristics the volume defines. The step in reasoning from volume to probability is not at all obvious. One powerful way of defining the probability of certain kinds of events is to form the ratio of the number of ways a particular event can occur to the total number of events that can occur. Thus the probability of flipping a coin and getting heads is 1/2, the ratio of the number of ways a coin can come up heads (one) to the total number of ways a coin could come up as the result of a flip (two). The probability of winning a lottery drawing in which a million differently numbered tickets have been sold is 1/1,000,000, the ratio of the number of ways your ticket can be chosen (one) to the total number of tickets that can be drawn (a million). The probability of rolling 7 with a single roll of two dice is 1/6, since there are six ways to get 7 (1 and 6, 2 and 5, 3 and 4, 4 and 3, 5 and 2, 6 and 1) and thirty-six possible combinations.

In Boltzmann's interpretation, the probability that a system will have the physical characteristics of a particular macrostate is the ratio of the number of ways the system might attain those characteristics to the total number of states the system could be in. By our definition, the number of ways a system could have the characteristics of a particular macrostate is the number of microstates in that macrostate, which is its volume in phase space. The total number of possible states of a system, regardless of its macrostate, is the total number of microstates it could have. This is the volume of the entire region of phase space that the system is permitted to inhabit. Since it is only macrostates that we can distinguish in our reality, we conclude that entropy inevitably increases. On the microscopic scale, however, the microstate can and does do anything it wishes, albeit while remaining almost always typical. It is guided by no laws, other than Newton's. The path of a system's microstate in phase space is extremely complicated, whereas the succession of macrostates that it enters is simple, reproducible, and irreversible.

The Ergodic Hypothesis

Could there be anything on the microscopic scale that would prevent a microstate from entering the high-entropy macrostate? If there were, then the process of coming to equilibrium that I have described would not happen. The "ergodic hypothesis," introduced by Boltzmann in 1871, states that a system's microstate will eventually move throughout all regions of phase space in accord with the physical constraints imposed by Newton's laws and the other fundamental laws of physics. For example, a system can't enter a region of phase space that represents more energy than it actually has. But aside from these requirements, the hypothesis asserts that a system's microstate can go anywhere and will. If true, it guarantees a microstate entry into the final, maximum-entropy macrostate.

The ergodic hypothesis was always assumed to be true, but as late as the 1950s nobody had attempted to prove or disprove it. In 1954, the Russian physicist A. N. Kolmogorov showed that large systems of vibrating atoms, like crystals, might not obey the hypothesis. His work prompted an intense study of such systems, assisted by the recent (at that time) development of digital computers. Through complex simulations, it was demonstrated that, indeed, such systems were not necessarily ergodic. The reason is that the atoms in a crystal tend to vibrate as a unit, called a "normal mode." Each of these normal modes is in a sense disconnected from all others. If the crystal is made to vibrate in a particular normal mode (we say that the mode is "excited"), it will stay in that mode and not become excited in any others. Thus the microstate of a vibrating crystal is really a description of which particular normal mode is excited rather than of what each atom is doing. Looked at in this way, it could be shown that the microstate of a crystal vibrating in a particular normal mode would remain forever within the region of phase space corresponding to that mode. The maximum-entropy macrostate, however, would correspond to a region in which all modes were vibrating. The crystal would never arrive at that macrostate.

We see this kind of behavior at the human social level also. We could think of a "social phase space" as defining all the interactions with others that a person could have. If a person moved freely through this space, touching all groups and individuals, we could think of it as a kind of ergodic existence. However, most people tend to stay within their own circles of friends and acquaintances, remaining in rather limited regions of social phase space as opposed to sampling the whole of it. They are not forced to limit their interactions by any immutable social laws, but rather by their jobs, their cultures, their neighborhoods, their economic circumstances, their intuitive feelings of "comfort," and the like.

It was somewhat distressing for scientists to find that a system as important and basic as a crystal did not obey the ergodic hypothesis. A way out was suggested. Suppose crystals actually had a physical mechanism by which different normal modes could communicate with each other. Such a mechanism would allow one normal mode to excite another and ultimately excite all modes. What could the mechanism be? In the language of physics, a small effect that connects otherwise unconnected phenomena is called a "perturbation." Perturbations arise in a crystal from imperfections that are invariably present. A few atoms out of place or a few atoms from another element are all it takes to perturb a crystal.

Unfortunately, perturbations did not accomplish what was hoped. Komogorov and two other physicists, Arnold and Moser, proved rigorously that these systems were still not ergodic. The so-called KAM theorem established a very interesting fact. Most of the paths of the crystal microstate remained in the restricted regions of phase space corresponding to the excitation of just a few normal modes. Perturbations did cause some paths to become extremely erratic and leave the restricted phase-space regions, but most of the paths remained trapped in their neighborhoods. Again, we see things occur in the social sphere that allow and even encourage people to move out of their immediate circles. Some do, but many do not. The effect is that there is more mixing within society, but we still see the remnants of social structures that have been only minimally affected by change.

Coming to Equilibrium

Following this probabilistic interpretation of mesoscopic reality, we can examine briefly another interesting phenomenon: the fact that a hot object and a cold object, when brought into contact, reach a state where they each have the same intermediate temperature, but the reverse process never occurs. By that I mean that if two objects initially at the same temperature are brought into contact, they are never observed to change their temperatures so that one becomes hotter and the other colder. It is with this certainty in mind that we place a pot of water on a stove if we want to heat the water. We expect the colder object, the water, to approach the temperature of the hotter object, the stove. We would certainly not place a pot of water in a refrigerator to heat it, in the expectation that the water will become hotter and the refrigerator colder.

Why does this happen? Let's examine the process from the standpoint of microstates and macrostates.

For simplicity I will assume that I have two objects, say two blocks of iron, of identical size and composition. One block is hot and the other is cold. The first block is hot because its molecules have more energy to share amongst themselves than do the molecules in the second block. At any instant, the energy in each block is distributed in a particular way among all its molecules. A block microstate corresponds to the way the energy is distributed among the molecules. For example, one rather unlikely (highly nontypical) microstate would have all of the energy residing in one molecule, with the remaining molecules totally devoid of energy. What kind of macrostate would such a microstate belong to? It would correspond to a block that was intensely cold, with the exception of one extraordinarily localized hot spot.

Another slightly different microstate might have one molecule with nearly all the energy and one other molecule with a minute amount of energy. There are many more microstates of this type than the previous type, because now we have the ability to move energy between two molecules. The situation corresponding to the greatest number of microstates is one where the energy is fairly evenly apportioned, so that every molecule has some energy. This situation allows more ways of moving energy among the molecules than if only a few molecules had all the energy and the rest had none. The actual physical process by which the energy in a block of iron moves about from molecule to molecule involves forces that the molecules exert on each other. Just as shuffling a deck of cards creates new microstates, the molecules move energy around by means of complex vibrations that affect each other. Thus the macrostate with the most microstates—that is, the maximum-entropy or equilibrium macrostate, is the one in which the energy is equally distributed among all the molecules in the block of iron. From a mesoscopic perspective, the block of iron in this macrostate would feel as though it were of uniform temperature. Whatever part of the block your finger happened to touch would contain molecules as energetic as those in any other part of the block. Since it is this energy that your skin interprets as "temperature," you would say that the block is uniformly hot or cold, depending upon which block you are touching.

When we place the uniformly cold block of iron, which is in equilibrium, in contact with the uniformly hot block of iron, which is also in equilibrium, we have in effect created a new system: a single block of iron, half of which is hot and half cold. Viewed now as a new system, this block of iron is not in its maximum-entropy macrostate because the energy within it is not uniformly distributed among its molecules. However, the contact between the two halves allows some of the excess energy in the hot half to move into the lower-energy cold half.

As I previously mentioned, this energy transfer occurs by a mechanism of vibrational interactions between neighboring molecules, and it now crosses the interface between the two halves. Without even considering the transfer mechanism in great detail, we can intuitively sense that the microstate of the single block will begin to move through phase space, changing progressively into microstates corresponding to a more uniform distribution of energy among the molecules. The microstate will eventually find itself within the maximum-entropy macrostate of the large block, where it will happily spend the rest of its life, with occasional fluctuations. Once the energy has distributed itself among all the molecules in the single large block, the energy per molecule is less than it was in the initially hot portion and more than it was in the initially cold portion. That is why the large block is found to have a temperature approximately halfway between the hot and the cold.

Microstates, Macrostates, and the Direction of Time

From a purely thermodynamic perspective, entropy must increase as a result of natural processes, so the direction of time can be interpreted in terms of the increase of entropy. We sense not time going forward, but rather entropy increasing.

From Boltzmann's microscopic point of view, however, entropy does not have to increase, it is simply most likely to increase; Although it is in principle possible to manipulate molecular velocities so as to make entropy decrease, such decreases will, with enormous probability, be temporary in nature.

One matter that puzzled Boltzmann greatly was the fact that, given how great the age of the universe must be, its entropy was still increasing. While that age had not been calculated, the age of Earth had been estimated by several important scientists, with Charles Darwin suggesting three hundred million years. It seemed reasonable that the microstate of the universe should have comfortably arrived at its maximum-entropy macrostate, and therefore, assuming it remained there, the entropy of the universe should be constant. Even if there were occasional fluctuations of the microstate out of the maximum-entropy macrostate, they would have to lead to slightly lower entropy macrostates and then return very quickly to the maximum-entropy macrostate. This implied that the universe might occasionally behave as though time were going backward, followed by time's return to its normal direction of movement. None of this seemed to be happening, however. The universe seemed quite content in its forward-time and increasing-entropy mode. Boltzmann finally arrived at the following explanation, which he wrote down in 1898:

> There must be in the universe, otherwise everywhere in thermal equilibrium,
> i.e., dead, relatively small regions here and there with the size of our star

space (let us call them individual worlds) that depart markedly from thermal equilibrium during the relatively short time of eons. Those individual worlds in which the probability of the state is increasing must be just as frequent as those for which it is on the decrease. (quoted in *Ludwig Boltzmann*, by E. Broda, trans. L. Gay and E. Broda [Woodbridge, Conn.: Ox Bow Press, 1983], 88)

Boltzmann thus concluded that we live in a portion of an otherwise "dead" universe that previously underwent a fluctuation to a lower-entropy state and is now returning to its final, highest-entropy state. He would concede the existence of other regions of the universe where quite the opposite is occurring. In those regions, the universe has suddenly embarked on a fluctuation toward lower entropy, and consequently physicists living on planets there would have discovered a second law of thermodynamics asserting the fact that, in all natural processes, the entropy of the universe must decrease!

Today we have a different view of things. What Boltzmann didn't realize, and what we now know to be true, is that the universe as a whole is expanding. Boltzmann, as well as virtually all the other scientists of his era, believed the universe to be static. An expanding universe continually gives its microstate room for movement without ending up in a terminal, maximum-entropy macrostate. Thus the universe is not yet dead, and we needn't explain increasing entropy by means of some local fluctuation that we happen to be living through.

The Ignorance Associated with Statistical Reality

How has our present understanding of statistical reality actually placed limits on our ability to understand the world around us? Prior to Boltzmann's investigations, we lived happily in a mesoscopic world in which we were certain of death, taxes, and other entropy increases. We knew that placing a pot of water on a stove would make the water boil and that placing an ice-cube tray filled with water in a cold enough place would produce ice cubes. Now, however, we have peered into another reality, the reality of teeming life on the microscopic scale. This is a reality that requires an enormous phase space for its description, not the ordinary three-dimensional space that suffices for describing our lives. In terms of this reality of the very many, we cannot view the processes of boiling and freezing as certain. We must view them in terms of probabilities rather than certainties.

Why is this probabilistic view needed? Couldn't we view the world of molecules in terms of Newtonian certainty and somehow recover the mesoscopic certainties of boiling and freezing? The answer, unfortunately, is "No." Certainties at the microscopic level apply to the behavior of microstates, but microstates have no existence in the mesoscopic world. There is no microstate that I could identify as a pot of boiling water. A microstate would be a collection of 1,000,000,000,000,000,000,000,000, . . . molecules, each with some amount of energy, bouncing madly into each other. The mesoscopic world is a world of macrostates, enormous collections of microstates that, taken together, form a recognizable entity. You and I see no microstates; we see only macrostates. To describe mesoscopic reality using the language of microscopic reality (which, by the way, we are not forced to do; it just happened that Boltzmann wanted to try it!), we must associate microstates with macrostates, and that involves probabilities.

In a very real sense, a macrostate is a confession of ignorance. It is a confession, however, that is necessary only because we now know that there are microstates. Had Boltzmann never begun his investigations, we could live in blissful ignorance of a molecular reality, as Mach insisted we do, and happily describe our existence with mesoscopic thermodynamics. But once we open our eyes to the presence of molecules and describe their behavior with microstates, we must also come to the realization that our world is made up of macrostates, which are large collections of microstates. Our ignorance, therefore, is our inability to discover which particular microstate we are part of. Does this really matter? Only on an intellectual plane, where we know there is knowledge that is beyond the abilities of our senses to obtain.

Entropy as a Measure of Disorder

As I have pointed out, the earliest definition of entropy, given by Clausius, involved only the mesoscopic thermodynamic concepts of heat and temperature and had no microscopic implications. It was Boltzmann who finally connected entropy, microstates, and macrostates and thereby gave entropy a "meaning" on the microscopic level as the multiplicity of a macrostate. Now multiplicity, the number of different microstates that correspond to the same macrostate, can be related to yet another concept, the "disorder" or "randomness" inherent in that macrostate. Disorder and randomness are somewhat intuitive and subjective concepts that are difficult to define precisely, but they are useful in helping us think about different forms of reality. The following analogy gives a sense of the connection between entropy and disorder.

Suppose I have ten molecules among which I intend to apportion five units of energy. Let's assume that the energy is distributed in integer amounts. I will define as a microstate of this system the set of ten numbers with an exact apportionment of the energy among individual molecules. For example, the set of numbers (4, 0, 1, 0, 0, 0, 0, 0, 0, 0) represents the microstate in which molecule 1 has four units of energy, molecule 2 has zero units, molecule 3 has one unit, and the remaining molecules have no units. A macrostate will simply be the distribution of units of energy, with no regard to which molecule has which unit. In this case, the macrostate is eight with zero units, one with four units, and one with one unit. The multiplicity of this macrostate is the number of different ways we can juggle the energies around to make different microstates. The four units could be anywhere among the ten molecules. Once a position for the four units is chosen, however, the one unit can be in only nine different positions. Finally, when we have positioned the four and the one, the remaining zeros are fixed. The total number of different arrangements is therefore (10×9), the number of positions of the four multiplied by the number of subsequent positions of the one. This gives a total of 90 different arrangements and therefore 90 microstates.

If we had put all five of the energy units in one molecule, that macrostate would have a multiplicity of 10. If we give five molecules one unit each, that macrostate has a multiplicity of $(10 \times 9 \times 8 \times 7 \times 6) = 30{,}240$. Emerging from all this is the fact that the most uniform distribution of energy gives the same amount to each molecule, and this macrostate has the highest multiplicity. It therefore has the highest entropy. This state is also the least "orderly," however, in the sense that you have the least knowledge of where the energy is. When a system is orderly, the states of its component parts are known with the greatest precision. If I know exactly which molecule has what amount of energy, I have the most precise knowledge possible of a system for which there are only two unknowns: what are the sizes of the energy packages and which molecule has them? If I know that the system is in a state in which one molecule has all five energy units and the others have none, then I am uncertain only as to which molecule has the energy. If I were to take a guess as to which molecule has the energy, I have a one in ten chance of being correct. When the energy is distributed among five molecules, I have only a one in 30,240 chance of guessing the distribution.

Thus high-entropy macrostates are highly disordered macrostates that leave me increasingly unsure which molecule has what energy. The more different values of energy that can be handed out, the higher will be the level of my ignorance. It is simply the fact that, the more there is to know, the more there is to be unsure of! Therefore we can restate the second law of thermodynamics to say that natural processes tend to drive the universe to increasing disorder, producing macrostates about which our knowledge is increasingly imperfect.

Entropy, Knowledge, and "Information"
By introducing the concept of knowledge, I seem to have injected a human attribute into this discussion of entropy. Certainly the universe goes to

states of higher entropy whether or not humans are around to attempt to gain knowledge of that process. Nevertheless, is it possible that human knowledge does play a role in defining the physical concept of entropy? After all, the human brain is also a physical system, albeit one that is extraordinarily complex and not understood to the degree that other physical systems are. What does knowledge correspond to in purely physical terms? Assuming it is stored in the brain, it must correspond to some "order" or pattern there. It may not be a visible pattern of cells, but it may be a pattern of electromagnetic waves or chemical concentrations or the like. In order to recall a thought, there must be an accessible degree of organization somewhere within the brain that can somehow be interpreted.

How might such a pattern be established? If the brain is ordinary matter then energy must be expended to establish a pattern where there was previously a different pattern. This energy involves the generation of heat through metabolic processes, the very kind of processes that the thermodynamic definition of entropy quantifies. We could even include the entropy that would be a necessary by-product of using electric lighting to read whatever books are required to increase our knowledge. Thus in principle at least we could assign a value to the entropy generated by the body and around it during the processes of gathering and storing new knowledge and even destroying old knowledge. This entropy would ultimately enter a person's surroundings and increase the entropy of the universe.

Suppose we consider, for a moment, some particular piece of knowledge, like the answer to a "true or false" question. If the answer is "true," I can assign that piece of knowledge a single number 1. If the answer is "false," I can assign it the number 0. When you hear the answer "False!" your brain, somewhere deep within its recesses, writes down the number 0 in the compartment it has set aside for that question. You and I might both hear that same answer, "False," and we will both write a 0 in our brains. But maybe I'm not quite as sharp as you are and my brain requires more entropy production to store and retrieve answers than yours does.

Scientists are not particularly interested in the amount of entropy your brain produces as compared to mine, but they are interested in the entropy associated with the structure of the answer itself. The simplest way to think about it is to imagine the answer written on the magnetic tape of a computer, rather than a brain. Computer tapes can be quantified much more easily than brains can. This whole branch of science is called "information theory," and the degree of order or pattern in the answer to a question—or in any statement, for that matter—is called its information content. If a statement is to have "meaning," the symbols that represent it must have a pattern. The greater the degree of pattern, the more the order and the less the entropy. Information is thus simply related to entropy, but it is applied to symbols such as letters and numbers rather than to physical processes. The letters and numbers require physical processes to produce them, store them, erase them, and of course use them, but the term "information" refers only to the pattern itself. In short, there will be a production of entropy due to the various processes that produce knowledge, but the knowledge itself, regarded as information, is a lowering of entropy, because it represents pattern where there was no pattern.

If you have studied biology or genetics, you learned that the double-stranded DNA molecule transmits from one generation to the next what are essentially the instructions for creating a human being. This miraculous molecule forms part of the manufacturing assembly line that produces the protein molecules used in the creation process. The instructions are carried as a series of four molecules that form a "code" for the synthesis of particular proteins. Each ordering of the four molecules along the DNA molecule causes the construction of a corresponding protein within the human cell. The proteins are produced at the right times and in the correct order to enable the cell to divide and ultimately form the proper human tissues and organs. Geneticists say that the DNA molecule carries information in exactly the way that a language does. The pattern of each specific series of molecules in the DNA has a measurable information content, which determines the complexity of the molecule that the DNA can in turn synthesize. The greater the information content, the more complex the possible synthesis.

The smallest unit of information is called a "bit," short for "binary digit," which is the amount of information necessary to answer a single "yes or no" question. On this sheet of paper, a 0 or a 1 is a bit of information, since it is the briefest symbol I could use to represent yes or no. On a computer tape, the orientation of a single magnetic region represents a bit. The particular symbol depends on the medium in which the message is written. The symbol acts as a trigger for some process that utilizes the information it carries.

As the human fetus develops in utero, it undergoes an extremely complex series of processes that increase its order. Cells that are initially similar gradually differentiate, becoming bone cells, muscle cells, skin cells, and so forth. Finally, these groups of cells organize still further and become tissues and organs. As we have already seen, when things are different and everything is in its place, it is a more orderly situation. One of the reasons the human fetus can develop in a manner that leads to increases in its order and seems to violate the second law of thermodynamics is that it contains within itself a reservoir of low entropy in the form of information, which guides its development.

On a somewhat lighter note, I like to consider the times I attempt to straighten my room. Spring-cleaning is a process that produces order from disorder and therefore decreases entropy. I put all my undergarments in one drawer, all my socks in another, and my pants and jackets on separate racks. Everything is in its place, and I have perfect knowledge of the whereabouts of my belongings. Does this process violate the second law? It cannot! Although I have decreased the entropy in my room, I have increased it in the universe around me by a greater amount. I have burned the electric lights, I have eaten and metabolized what I ate, I have perspired, I have thrown out trash. All of these activities have added to the store of entropy that surrounds me. I can lower my personal entropy and thereby improve my state of well-being only at the expense of my neighbor's entropy.

Well-being is to a large degree associated with low entropy. Life is a process of maintaining a low state of entropy in a universe that is inexorably increasing its entropy. The reason that Earth survives is that it gets heat and light from the Sun. This in turn is a result of the fact that the Sun is very hot while Earth is very cold, which is an ordered, low-entropy situation. Eventually the Sun will cool and Earth will warm. This is a higher-entropy situation and the one toward which the universe is tending. Survival depends on differences, which are evidence of low entropy. When the universe evolves to its final state of high entropy, all differences will have been eliminated. Instead of being made up of large objects at different temperatures, separated by huge regions of emptiness, the universe will be uniformly filled with dust at a constant temperature. That is the equilibrium macrostate of the universe, a situation that can support no life.

In a political and social sense also, meaningful activity is driven by differences. Differences in social and financial levels drive people to improve themselves. Differences in political power drive people to achieve power by organizing. Organizations themselves are evidence of low entropy and are capable of effecting change. On the other hand, if power is uniformly distributed among individuals in the way that energy is ultimately distributed among the molecules of a block of iron, it is evidence of an approaching equilibrium and a more static situation. If one could define a social entropy analogous to thermodynamic entropy, it would seem that democracy was a higher entropy state than, for example, monarchy. I make no judgment of "good" or "bad" for low- and high-entropy social systems, nor can I prove a second law of "sociodynamics" that is analogous to the second law of thermodynamics. In short, I don't know if it is a natural social tendency to go to states of disorder from those of order. Nevertheless, it is interesting and sometimes fruitful to apply the methods of analysis that work for one discipline to the problems of another.

Chaos and Irreversibility

There is one aspect of irreversibility that I rather neatly glossed over in the previous discussion. Boltzmann's conception depends critically on the typicality of a microstate. Typical microstates do predictable, describable things. Untypical microstates could decrease entropy, but they don't remain untypical for very long. The question is, Why are untypical microstates so unstable? What makes them revert to typicality at the slightest provocation? The answer to this question may involve the mathematical phenomenon called chaos, which is important enough to merit its own chapter later in this book. Without giving away the plot too early, let me simply say that chaotic behavior cannot be distinguished outwardly from randomness, yet at its core it is deterministic. This behavior very likely drives the molecular motion that Boltzmann described. It probably explains why his assumption that collisions randomize motion is actually true, even though the reasoning goes beyond what Boltzmann knew in his time.

Ten

From Classical to Quantum Physics

The universal acceptance of the power of scientific explanation as embodied in the methods of Newtonian physics was quite understandable. After all, Newton's mechanics, grounded in both mathematical analysis and empirical observation, could explain and predict with unprecedented precision motions as disparate as the trajectories of cannonballs and the orbits of planets. In a similar manner, the science of thermodynamics, systematically developed by Carnot, Joule, Clausius, and Kelvin, could describe and predict virtually all of the observed thermal behavior of gases, liquids, and solids. The highly theoretical statistical mechanics developed by Boltzmann, while not yet as widely accepted as the more empirical thermodynamics, was at least the beginning of an attempt to explain thermal phenomena in terms of the underlying atomic and molecular structure of matter. Boltzmann's invisible molecules behaved in accordance with Newton's laws, so that even at this unseen level of reality there was still strict adherence to classical laws, just as Copernicus's break with Ptolemy retained Plato's perfect circular orbits.

Maxwell's equations of electrodynamics seemed more than adequate to describe and explain all known electrical and magnetic phenomena. Moreover, these equations established the surprising and exciting fact that electricity, magnetism, and light were to be considered not as separate and distinct phenomena but as different manifestations of a single unified concept, the electromagnetic field. Being able to demonstrate that all the properties of light, particularly the precise value of its speed, almost miraculously emerged from a set of equations and experiments that at no point made any reference to light itself reportedly moved Maxwell to exclaim, "The only use made of light in the experiment was to see the instruments."

Causality, Determinism, and Predictability

Taken together, the experimental-theoretical systems of mechanics, thermodynamics, and electrodynamics provided as detailed, useful, and practical a map of macroscopic reality as any scientist could ask for. Anyone using this map could not help but be convinced that the world itself was as causal, deterministic, and predictable as the map.

By the term "causal," I mean that each of these theories explained the occurrence of phenomena in terms of conditions that preceded them. There were no miracles allowed in these theories, no intervention of the gods, no acts of chaos or randomness. Every effect had a preceding cause, and every cause produced an effect. And most important, both the causes and the effects could be described and explained within the context of the theories. In mechanics, for example, forces were the causes and accelerations were the effects. Planetary orbits and cannonball trajectories were manifestations of these causes and effects, being simply the paths of accelerating material objects that were being acted upon by external forces such as gravity.

With the term "deterministic," I refer to a property of the equations that gave mathematical expression to the theories. An equation is called "deterministic" if a given set of initial values of its variables corresponds to a unique solution of it. If you know the values of the variables at any given time, you know them for all time, Thus if Newton's equation of motion is solved for the trajectory of a cannonball, it is necessary only to know where the cannon is situated and with what velocity the cannonball emerges from its muzzle in order to know precisely where it will land. That is the essence of mathematical determinism. The cannonball must move along a unique path, because that alone is what the equation permits.

The final term, "predictable," refers to the ability of the theories to predict the outcomes of experiments and the course of natural phenomena. You might wonder why this property is worthy of having its own name rather than being just another aspect of determinism. The concepts are related but not the same. Predictability implies the ability to act on determinism. We shall see a bit farther on, when we discuss chaos, that special circumstances can render a phenomenon unpredictable, even though it is described by equations that are mathematically deterministic.

It is comforting to have a set of theories that are causal, deterministic, and predictable, because those three attributes are already a part of our genetically derived intuitive sense of the nature of the world around us. When the theories we devise reinforce our intuition and common sense, they become infinitely more acceptable. So when physics produces a map of macroscopic reality that is completely in tune with our expectations of that reality, we grasp the map with confidence and begin to use it. Indeed, it is doubtful that the user of such a map even notices the map! Surely it is the terrain itself, the real world, that possesses these attributes. The map simply represents them in a particularly convenient mathematical form. It is like a sheet of transparent plastic through which we look directly at the world that lies before us.

Let me clarify the meaning of my map analogy in this context. The theoretical constructions of Newton, Maxwell, and the various developers of thermodynamics were all expressed in mathematical terms. Thus the map of the physical world is expressed in the language of equations. The equations, however, describe not the actual behavior of physical objects but rather the mathematical behavior of certain symbolic representations of objects. If it were your intention to solve Newton's equation for the motion of the Moon around Earth, the symbols of interest would be $x(t)$, $y(t)$, and $z(t)$, the three coordinates of the center of the Moon at the time, t, of observation. Thus the solution is a set of three dependent variables, x, y, and z, expressed in terms of a fourth independent one, t. We make the assumption that the three position symbols *are* the Moon itself. There is an unstated assumption that if I went outside at night and looked at the Moon, it would be located at precisely those coordinates. I don't have to keep looking at the Moon to make sure that the coordinates aren't lying to me. The equation works whether or not I am actually observing the Moon. Moreover, it goes without saying that my act of looking at the Moon does not affect the course of its motion. The Moon, I assume, has an objective reality that goes far beyond my desire to describe its motion.

We say, therefore, that the map describes an independent, external reality, which is there whether we choose to describe it or not. My equations are describing "the Moon," not "my perception of the Moon" or "the way the Moon behaves because of the fact that I am describing it." We don't expect a map of the Sahara Desert to contain descriptions of the various mirages that travelers sometimes see. Those mirages may regularly appear in certain locations, and they may be disturbingly real to those who observe them, but they do not constitute the independent, external reality of the Sahara Desert. It may be argued that a map of mirages could be of great use to a traveler, but such a map should be marketed as a map of mirages, not as a map of the Sahara Desert.

This leads to a good question: how do we separate independent reality from the mirages brought about by our own states of mind? How do we know what belongs in the map and what does not? In mesoscopic/macroscopic reality, the answer is simple. You do experiments that separate mirage from fact. If you jump into the pool at the oasis but land in a pile of sand, you delete that oasis from the map. This all seems rather straightforward. We shall soon see, however, that in our attempts to describe the microscopic world, the distinctions between external reality and the symbols on the map that are part of the mathematical description of that reality become blurred and confused. So much so, that even though nearly a century has elapsed since the beginnings of studies of microscopic reality, serious arguments persist over the interpretation of the maps and how they should be used. When brilliant minds disagree with such vehemence and for so long, you can be assured that microscopic reality is *very* different from mesoscopic reality. In fact, much of the difficulty in understanding microscopic reality stems from our expectations that microscopic reality should behave just like mesoscopic reality and be subject to the same type of description. We have come to expect the same certainties, the same degree of precision in experimental results, and the same clarity of description that we have grown used to in dealing with the mesoscopic world. It is our right, is it not? Unfortunately, it is not. I will try to give the flavor of these arguments and some insights into the reasons they exist, but don't expect me to provide definitive answers to questions about the nature of the microscopic world. That would spoil all the fun.

A Revolution in Mapmaking

If there is an identifiable beginning to the end of the universal applicability of classical physics, a time when belief in the traditional maps of macroscopic reality was first shaken, it is at the very outset of the twentieth century. The first hints that something was amiss were very subtle. Many experiments were carried out during this period, all of whose results failed to be adequately explained by the theories of classical physics. We have already discussed the Michelson-Morley experiment, which seemed to indicate that Earth moved at no measurable speed through the ether. Science initially responds to unexpected observations and experimental results by attempting to modify existing theories in as modest ways as possible. In the case of Michelson-Morley, there were valiant efforts by physicists like Fitzgerald to retain the ether. In a similar manner, the experiments I will now discuss were met with the assumption that there was nothing seriously wrong with classical physics, an assumption that turned out to be unwarranted.

These experiments and observations were (and still are) in clear conflict with the accepted theories of macroscopic reality. I review these selected historical developments only to bring you up to date on the current state of affairs. I am not a historian of science, and my purpose is not to present anything like a complete history of the demise of the universal validity of classical physics. Here then is a list of six experiments and observations that, as much as any others, played important roles in that demise. Another author could easily come up with an alternative list.

1. The classically inexplicable stability of Rutherford's atom (1912)
2. The discrete spectra of excited atoms observed by J. J. Balmer (1885)
3. Blackbody radiation (1890s)
4. The photoelectric effect (1905)
5. The electron diffraction experiment of Davisson and Germer (1927)
6. The complementary nature of observations exemplified by the experiment of Stern and Gerlach (1922)

These observations and experiments can be viewed as pieces of a puzzle that, when finally put together, gave evidence of a reality so different from the one we know and requiring a mode of description so unusual that not even Einstein could believe it.

The Stability of Atoms

Items 1 and 2 belong together, so I will discuss them as a pair. Maxwell predicted that accelerating charges would produce electromagnetic radiation, which in turn would carry energy. Ernest Rutherford's theoretical model of the atom envisioned it as a miniature solar system, with negatively charged electrons rotating around a positively charged nucleus, just as the planets rotate around the Sun. Rutherford's theory was verified by an experiment carried out by Hans Geiger and Ernest Marsden in which certain charged particles (alpha particles) were fired at a sheet of gold foil. While most particles passed through the foil, indicating that the foil was primarily devoid of matter, some of the particles bounced back, indicating the presence of small, extremely massive objects scattered throughout. These were the atomic nuclei. According to Newtonian mechanics, electrons rotating around nuclei are accelerating, and therefore, by Maxwell's theory, they should radiate electromagnetic energy. That radiation in turn should diminish their mechanical energy, so the electrons should ultimately spiral into the nuclei of the atoms. If this were the case, all atoms would eventually cease to exist. Clearly it was not the case; atoms did in fact exist (Mach and his positivist philosophy by now having been decisively refuted).

Atoms did radiate; they just didn't radiate continuously, and they certainly didn't radiate themselves into oblivion. It was well known at the time that very hot gases radiated visible light. In 1885, the physicist J. J. Balmer showed that when hydrogen gas was "excited," meaning that it was given extra energy by either heating it or passing an electric current through it, it gave off visible light, presumably because it was expelling the excess energy. This light, however, was composed of just four frequencies or colors, which, taken together, are called the visible spectrum of hydrogen. These four components had remarkably precise wavelengths, which Balmer measured and found to be 656 nm, 486 nm, 434 nm, and 410 nm. (The symbol nm is short for "nanometer," 10^{-9} meters, a billionth of a meter.)

The problem with this situation was twofold: 1) the definite frequencies of the light observed by Balmer weren't consistent with the continuous spectrum of radiation that electrons spiraling into the nucleus should be emitting, and 2) the light was given off only by excited atoms and not by atoms in their normal, unexcited states, whose orbiting and accelerating electrons should also be constantly radiating. The confusing situation implied inconsistencies in Rutherford's model, Maxwell's equations, or Newton's mechanics, separately or in combination.

Blackbody Radiation

The phenomena of blackbody radiation and the photoelectric effect, items 3 and 4 in my list, also involved unexpected properties of the interaction of radiation and matter. In these phenomena, however, it was the amount of radiation, its intensity, that created difficulties.

Blackbody radiation is a strange name for what, to anyone but a physicist, is a mundane and unexciting phenomenon, the electromagnetic radiation given off by any finite object maintained at a definite temperature. The adjective "blackbody" derives from the fact that when heated objects are to be studied, they are painted black so as to eliminate the extraneous effects of reflected light. Blackbody radiation includes the glow that emerges from the interior of a heated hollow object, such as a kitchen oven, through a small window in it. For this reason, it is sometimes called "cavity" radiation.

Physicists thought they could explain this type of radiation in terms of classical concepts. As Balmer demonstrated in the case of hydrogen, if you excite a single molecule or a diffuse gas of individual molecules, the resulting radiation comes out as a definite set of frequencies and wavelengths called a "discrete spectrum." If, however, you heat a large solid object composed of huge numbers of atoms held tightly together by interatomic forces, the radiation produced loses the individuality possessed by the radiation of individual atoms. The radiation is now mostly produced by collective vibrations of the solid rather than by vibrations within the individual atoms and molecules that compose the solid. The discrete spectrum of the individual atoms is replaced by a continuous spectrum associated with the solid, the blackbody spectrum, whose

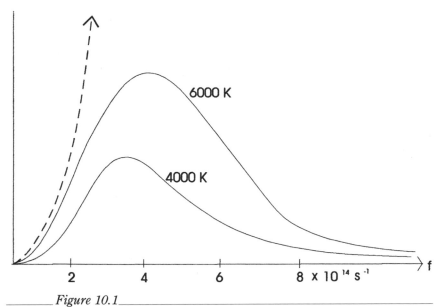

_____ *Figure 10.1*_____
Blackbody Spectrum and Ultraviolet Catastrophe

characteristic shape depends on the temperature of the material and not its composition.

The spectrum of blackbody radiation had been extensively studied by the physicists Otto Lummer and Ernst Pringsheim, and its shape had been experimentally determined. By the "shape" of a spectrum, I mean the intensity of the different frequencies that compose it, the brightness of its colors. Figure 10.1 is a schematic representation of the blackbody spectrum for two temperatures, 4,000 kelvins and 6,000 kelvins, the latter being the approximate surface temperature of the Sun. I have also sketched in a third curve, parabolic in shape, which we shall see is the spectral shape predicted by classical analysis.

The blackbody spectrum has several interesting features, none of which seem unduly worrisome. First, there is a distinct peak in the spectrum, where the frequencies of radiation are most intense. The location of this peak depends only on the temperature of the blackbody; the hotter it is, the more the peak moves toward higher frequencies (a phenomenon called Wien's displacement law). This is why objects begin to glow as they are heated and why the color of that glow changes from red to orange to yellow as the objects become hotter. A "red-hot" object is one whose temperature corresponds to a spectral peak in the frequency range of visible red light. All blackbodies radiate at all frequencies, but the peak dominates the spectrum, and if the peak happens to be in the visible region, it is that color that we see.

The second feature of interest is the fact that the spectrum tapers off at both low and high frequencies. Low-frequency radiation includes such things as radio waves, microwaves, and infrared radiation. High-frequency radiation includes ultraviolet radiation and X rays. We cannot see either the high or low end of the spectrum, but each is there nevertheless. The Sun is a blackbody whose spectral peak is in the visible yellow portion of the spectrum, but it emits a good deal of infrared radiation, which heats us, and ultraviolet radiation, which produces sunburn, not to mention radio waves and X rays. Much of the energy in these frequencies is absorbed in our atmosphere, so we tend to ignore them.

What is it about blackbody radiation that ran counter to the map of classical, macroscopic physics? It won't seem like much, but its implications are enormous. Basically, it is the fact that there is far too little high-frequency radiation. Classical physics predicts a spectrum that looks like the dashed curve in figure 10.1, in accordance with the so-called law of equipartition of energy. This law asserts that, when a system is given a certain amount of energy and arrives at a constant and uniform temperature, the energy will be equally apportioned among all of the "degrees of freedom" of the system, the types of motion or behavior the system can have. If you can jump up and down and also turn cartwheels, each of those motions is one of your degrees of freedom, something you can do independently of all the other things you can do. According to the equipartition law, if I give you ten units of energy, five will go into making you jump up and down and the remaining five into having you turn cartwheels. The more things you can do, the greater the number of degrees of freedom you have and, as a consequence, the more finely divided will be any energy you are given.

The concept of energy equipartition is related to our previous discussion of maximum entropy and the approach to equilibrium. Every particular distribution of energy among degrees of freedom constitutes a microstate of a system. As the system evolves in time, its energy is transferred between different degrees of freedom until a maximum-entropy macrostate is reached. This macrostate will correspond to a fairly uniform distribution of energy among all the degrees of freedom of the system. Equipartition of energy, in short, is another consequence of the approach to equilibrium of macroscopic systems. Just as a gas moves so as to uniformly fill the container it is in, so does energy move so as to uniformly "fill up" degrees of freedom.

What then are the degrees of freedom for blackbody radiation, and how does equipartition of energy affect them? The answer is not particularly simple. In fact, for a long time the best physicists in the world came up with the wrong answer.

First let's define the system. If we take as the blackbody a large cavity filled with radiation, like an oven, then the system can be thought of as either the oven itself or the radiation that fills it. The usual treatment of blackbody radiation takes the radiation itself to be the system. The oven is of course producing the radiation and providing the energy to sustain it, but once that radiation fills the oven, it can be treated as an independent entity. Think of an orchestra in a concert hall producing a sudden crescendo of sound and then ceasing to play. As soon as the sound is produced, it has a life of its own. It bounces back and forth between the walls, creating echoes and variations in intensity. Finally its energy is absorbed by the audience and the walls, and it disappears.

We can think of the radiation produced by the vibrating molecules in the material of the oven walls in an analogous way. The radiation is produced, fills the interior of the oven, and dies away when the oven is turned off. However, we will not turn the oven off. We will allow it to reach and remain at a temperature of T kelvins. The radiation in the oven will therefore persist at a corresponding level of intensity. You might wonder why the radiation doesn't keep increasing in intensity if the walls of the oven keep pumping it out. The reason is that the radiation flows back into the oven walls if it becomes too intense. We say that the radiation and the oven reach a state of thermal equilibrium in which both are at the same temperature. In this state, every additional amount of emitted energy is balanced by a similar amount of reabsorbed energy.

It seems odd to think of radiation as having a temperature. We can easily imagine matter as having a temperature, but it is difficult to think of light or radio waves that way. What we mean is that the radiation will be in equilibrium with matter at the temperature at which it is produced. The radiation will neither increase nor decrease in its total energy content.

When we look into the interior of an oven, therefore, we are observing radiation in equilibrium with the oven at the oven's temperature. That is what we mean by blackbody radiation. The problem at hand is to determine why its classically predicted characteristics are completely at odds with its observed characteristics.

What Is Radiation?

Radiation, of whatever kind, is actually a fairly simple phenomenon to describe. First, we must understand that radiation is the vibration of something, be it a material medium (like air in sound radiation) or the field in electromagnetic radiation. As a vibration, radiation has three important characteristics: frequency, amplitude, and speed of propagation. Frequency is the number of times per second that the radiation vibrates, amplitude is the magnitude or height of vibration, and speed of propagation is the velocity with which it moves from place to place. Because radiation does move, the frequency of its oscillatory motion results in a wavelength, the distance it moves in the time it takes to perform a complete oscillation.

In sound radiation, which is not electromagnetic in origin but shares the same basic features, the frequency corresponds to what we call the "pitch" of the sound, while the amplitude is closely related to its loudness. In visible light, which is electromagnetic radiation, frequency corresponds to color and amplitude to brightness. Other kinds of electromagnetic radiation, like X rays or microwaves, are not directly perceived by our senses, so we have never developed words to describe our impressions of their frequency and amplitude. But the concept is always the same: rate and magnitude of vibration.

Since blackbody radiation is produced by vibrating molecules, it is composed entirely of the frequencies at which the molecules vibrate, singly and in combination. It is not radiation of a single frequency, but rather a composite of an enormous number of different frequencies and amplitudes, which is why it is described in terms of a spectrum. In referring to the sound in a concert hall, we talk of an orchestral spectrum, because musical instruments produce sounds of different frequencies.

According to classical physics, vibrating molecules can produce virtually any frequency and its corresponding wavelength. The cavity containing the radiation, however, will not accept or sustain any wavelength, but only those that fit precisely within its walls. This important restriction severely limits which long wavelengths will occur within the cavity. A similar effect occurs with sound. You will not properly hear the lowest notes of an organ in a concert hall whose dimensions do not contain at least half a wavelength of those notes. The wavelength of a 16-cycle-per-second organ note is 68.75 feet. It takes the sound vibration 1/16 of a second to complete a cycle, and with the speed of sound being approximately 1,100 feet per second, the wave travels (1,100/16 = 68.75) feet in that time. If the room cannot hold a wave of this size, the vibration doesn't establish itself effectively and is relatively inaudible. What you will hear instead is the next "harmonic" of the organ note, which has a frequency of 32 cycles per second and a wavelength of 34 feet. Although an instrument can produce a sound, that doesn't mean that the sound will be sustained by its acoustical environment.

The higher the frequency of the wave, the shorter is its wavelength and the more easily it fits into a cavity of a given size. There is no difficulty hearing a piccolo in a telephone booth (its wavelengths are only a few centimeters) or fitting visible light into the smallest oven. Recall Balmer's demonstration that the visible light produced by hydrogen has a wavelength of only millionths of a meter. Even the smallest microwave oven will find space to hold such radiation.

The effect of this "fitting" phenomenon is that the blackbody radiation in a cavity should contain much more high-frequency radiation than low-frequency radiation, not of course because that is what the oven walls produce, but because that is what the oven cavity can accept. A microstate of the radiation in the cavity refers to the relative amounts of radiation at different frequencies. According to the equipartition law, each bit of radiation with a different frequency (called a "mode") is entitled to an equal amount of energy. The frequency of a mode is a degree of freedom in blackbody radiation. Since there are so many more high frequencies than low frequencies, however, the preponderance of available energy should be found in high frequencies. Therefore the spectrum of blackbody radiation should be overwhelmingly powerful at its high-frequency end. This prediction was somewhat humorously called the "ultraviolet catastrophe" (see fig. 10.1), since ultraviolet radiation represents the high-frequency end of the visible spectrum. Classical analysis also predicted an infinite amount of energy in the blackbody spectrum, which was an additional impossibility. Fortunately for common sense but unfortunately for classical physics, there was neither an ultraviolet catastrophe nor an infinite energy content in the spectrum.

In fact, the high-frequency end of the blackbody spectrum goes rather benignly to zero, indicating that very little energy resides in these high frequencies. This difference between theory and actuality is about as great a discrepancy as classical physics ever produced. Coming as it did when classical physics was at the pinnacle of its successes, it offered a somewhat bitter taste of reality. Pride had to be swallowed and an explanation provided. The development of that explanation formed the foundation of modern physics, once and for all establishing the sad fact that classical physics had limitations. There was more than a roast in that oven. Looking into that oven, plain and simple, was looking at a new reality!

The Photoelectric Effect

The photoelectric effect is a phenomenon in which light shining on certain materials causes them to emit electrons. Like blackbody radiation, the effect had a rather simple classical explanation. Unfortunately, also like the case of blackbody radiation, the explanation was inconsistent with observations.

The classical interpretation was straightforward. Just as the molecules in an oven wall vibrate because of their mechanical energy and thereby produce electromagnetic radiation (their charged electrons vibrate along with them), so the inverse process also occurs. Radiation impinging on solid materials causes their electrons to vibrate, imparting sufficient mechanical energy to the electrons to literally shake them loose from their atomic structure and eject them into the external world. Based on this interpretation, it could be shown that the energy with which electrons were ejected from their parent material would depend on the amplitude (the intensity or brightness) of the radiation. Electrons would not necessarily be ejected as soon as the radiation struck them, since there was a certain minimum energy (called the "work function") needed to overcome the attractive forces produced by the surface of the material. The amount of time it would take for electrons to be ejected should therefore also depend on the energy of the radiation. Incident light that was insufficiently energetic would have to shine on the material for a longer period of time for enough of its energy to accumulate in the electrons, overcome the work function needed to cross the surface, and eject the electrons from the material.

The classical model of the photoelectric effect was shown to be not what actually happened. First of all, the energy of the ejected electrons didn't depend at all on the brightness of the incident radiation; it depended only on the radiation's frequency (the color). Furthermore, electrons were ejected immediately, even if the radiation was insufficiently bright, when the frequency was above a certain value. If the frequency was below this value, electrons were not ejected at all, regardless of the length of time the radiation was incident on the material. Finally, increasing the brightness of light of a given frequency had no effect on the energy of the emitted electrons; the only change was to increase the number of emitted electrons.

Resolution through Revolution

Neither blackbody radiation and the photoelectric effect nor the stability of atoms could be explained by the classical theories of physics. Attempts to modify the theories slightly in order to bring them into line with experimental facts did not improve the situation measurably. Resolution would come only through radical change, through revolution rather than evolution.

The discrepancies between the observed blackbody radiation spectrum and its classically predicted counterpart were resolved by a bold stroke of genius in 1900 by the physicist Max Planck. In an equally brilliant manner, the questions about the photoelectric effect were resolved by Albert Einstein in 1905. The two scientists' proposals involved very similar hypotheses. The stability of the atom was explained in 1913 by Niels Bohr with an equally revolutionary hypothesis and an entirely new way of envisioning the atom. These three scientific revolutionaries opened up a window on the strange world of micro-scopic reality.

Planck's Hypothesis and Blackbody Radiation

Planck could find no fault with the classical prediction of the number of different frequencies of radiation that could fit into a heated cavity or with the equipartition law. The only way to explain the lack of energy in the higher frequencies was to hypothesize that the energy offered under the equipartition law could not always be accepted by the radiation. In short, electromagnetic radiation of a given frequency would not accept energy in an arbitrary amount from the material of the cavity. There would have to be a certain minimum amount of energy available before the radiation would accept and use it.

In order to derive a spectrum whose shape corresponded to the observed one, Planck had to calculate this minimally acceptable energy. After much la-borious calculation, he concluded that an electromagnetic vibration of fre-quency f would have to accept energy only in "lumps" that were integer multiples of a basic size hf, where h was a constant he determined to be 6.6256 $\times 10^{-34}$ joule-seconds, now called "Planck's constant." If energy a bit more than hf but less than $2hf$ was offered, only hf would be accepted. If energy a bit less than hf was offered, nothing would be taken. The effect of this hypothesis was that, at any given temperature, the energy in an oven would be equally divided among all the available frequencies, but the divided amount would be less than some very high frequency was willing to accept. All the frequencies above that frequency would also find the offered amount insufficient and would not be energized. Instead of the high frequencies grabbing the lion's share of the avail-able energy and dominating the spectrum, there would actually be a maximum frequency above which no energy would be found.

We can now see why the spectrum was devoid of energy at both the high and low frequencies. At low frequencies there was no energy because these frequencies simply weren't there; they had wavelengths that were too large to fit into the cavity. This was classical reasoning, and it was correct. At high fre-quencies there were many wavelengths willing to play the game, but they were too greedy and demanded more energy than was allotted to them by the law of equipartition. This was Planck's reasoning, which was not classical but was also correct.

A fundamental question remained: why did radiation of a given frequency f demand its energy in lumps of exactly hf? That question has no answer other than "That's the nature of reality at the microscopic level." Planck's hypoth-esis was not meant to explain why something happened; it was meant to ex-plain how it happened. In a sense, it is a reversion to the Platonic mode of explanation, where only the gods know why and mere mortals must be content to describe how. We shall see, however, that the fact that microscopic reality does behave this way has far-reaching consequences for the way it must be described.

Einstein's Hypothesis and the Photoelectric Effect

Five years later, Einstein devised a similar hypothesis to explain the photoelectric effect: the electrons in the emitting material would accept energy from light only in the same lumps of energy, hf, that Planck had hypothesized were accepted by radiation. Whereas Planck saw this as a restriction on the energy that could be transferred from matter to radiation, Einstein used it as a restriction on the amount of energy that could be transferred from radiation to matter. Einstein called the lumps of energy "quanta," and that name has persisted to this day.

Einstein's explanation went briefly as follows. When light of a given frequency f struck an electron, a quantum of energy, hf, would be transferred in an "all-or-nothing" exchange. The amplitude or brightness of the light had nothing to do with the amount of energy it could give to an electron. The amplitude, according to Einstein, determined how many quanta per second were carried by the radiation, not the size of an individual quantum. If an electron required a ten-unit quantum to leap from its material confinement, the most intense light made up of nine-unit quanta would not do it a bit of good. A nine-unit quantum would be unacceptable to the electron. On the other hand, even a dim light of a few eleven-unit quanta would immediately cause the electron to be ejected, with one unit of energy to spare. An intense beam of ten-unit quanta would eject electrons profusely.

The Quantum of Knowledge

If you combine Planck's hypothesis with Einstein's, you must conclude that matter and radiation can interact only by the exchange of quanta of energy. Matter is unaffected by light-energy quanta that are too small to be acceptable, and light of frequency f will not be produced by energy in matter that is less than hf. There appears to be a fundamental "graininess" in the interaction between matter and radiation. Moreover, this graininess is as intrinsic to radiation as it is to matter. Not that light's energy is delivered in lumps because that is what matter finds acceptable. Einstein himself is said to have once remarked, "Although beer is sold only in pint bottles, it does not follow that it exists only in indivisible pint portions." Unlike beer, however, the energy in electromagnetic radiation does seem to exist in quantum portions.

We shall soon see that this graininess does much more than contribute to unusual phenomena such as blackbody radiation and the photoelectric effect; it also places fundamental limits on the study of microscopic reality and its behavior. As the interaction between matter and radiation becomes grainy, the nature of knowledge and the way we acquire it become uncertain.

Explaining Atomic Stability: The Bohr Model of the Atom

In 1916, Niels Bohr answered the two-part question of why atoms are normally stable and why they radiate only with discrete frequencies when they are excited. Bohr's answer changed the model of the atom from Rutherford's tiny solar system to a version that had no classical counterpart. It was a complete break from classical physics because it involved assumptions that were classically unthinkable.

Bohr attacked the problem of the atom's stability with the philosophy if it didn't happen, it couldn't happen. The absence of evidence that electrons were spiraling in toward the nucleus of an atom led Bohr to propose that spiral paths were forbidden and that electrons must normally be locked into orbits at a fixed distance from the nucleus. In those orbits, atomic electrons would have a constant energy and, contrary to classical beliefs, would not radiate.

To explain or at least justify these fixed orbits, Bohr borrowed Einstein's terminology and proposed that certain physical attributes of orbiting electrons were "quantized." In this context, Bohr meant that the values of these physical attributes could only be multiples of a certain constant, Planck's constant to be precise. Thus electrons in an atom must occupy an orbit characterized by one of these values. In Einstein's treatment of the photoelectric effect, the quanta referred to units of energy. In Bohr's treatment of the atom, quanta referred to units of orbital angular momentum, a physical property associated jointly with the radius of an orbit and the speed with which an electron traversed the orbit. Because its orbital angular momentum was quantized, the electron was forced to inhabit a particular set of orbits, determined by their distance from the nucleus. This in turn allowed the electron to have only a special set of possible energies.

Bohr then combined his hypothesis of quantized orbits with Planck's hypothesis that matter transfers energy to radiation only in units of hf, asserting that when an electron is in one quantized orbit, it can move to another only by emitting radiation whose frequency, f, is determined by $f = \Delta E/h$, where ΔE is the energy difference between the two orbits. Instead of the continuous emission of radiation predicted by classical physics, there is the discrete emission as observed by Balmer, with frequencies determined by the energy difference between allowed orbits. Maxwell would have electrons moving down slides, emitting radiation continuously as they go. Bohr has them going down stairs, emitting radiation only as they take a step downward and absorbing radiation that allowed them to take steps upward. In Bohr's model there is a final, bottom step, called the "ground state," below which an electron cannot go. It corresponds to the closest position to the nucleus and to the least energy for the electron. This ground state is the normal state of an atom, in which it can live for an eternity.

Matter as Waves: De Broglie's Hypothesis and the Davisson-Germer Experiment

Each of the observed phenomena I have been discussing ultimately required for its explanation the formulation of radical new ideas. That is the usual sequence of affairs in science: first observation, then explanation. I will now deal with the opposite situation, an observation that substantiated an idea. The idea was put forward by Louis de Broglie in an attempt to make the new atomic theory as complete and all-encompassing as possible. It was perhaps the boldest idea of all.

De Broglie proposed the following. If, as Einstein hypothesized, radiation interacts with matter by depositing its energy in discrete units, then we are ascribing to it a distinctly particulate aspect. This is an entirely new concept. Radiation had previously been regarded as the epitome of a "continuous" phenomenon. Its interaction with matter occurred smoothly, without breaks or hesitations. Now scientists were being asked to believe that this was not so, that there was a definite lumpiness or graininess in its character. Such lumpiness had always been the exclusive property of the things we call particles.

If light, which is a wave, can behave like a particle, then why can't the opposite also occur: why can't a particle behave like a wave? This was de Broglie's radical suggestion: that all particles should have wavelike properties. In particular, he proposed that *every* particle should be associated with a wave and a wavelength determined by the classical momentum of the particle. De Broglie presented the following deceptively simple relationship between the wavelength, λ, and the momentum, mv, of the particle:

$$\lambda = h/mv.$$

In words, this equation says that the wavelength associated with a particle is inversely proportional to its ordinary momentum, with the constant of proportionality being, yet again, Planck's constant.

In principle, de Broglie's equation applies to everything from an electron to a planet. We are all particles, solid, finite in size and having definite boundaries. That's what I mean by being a particle. We can see from the equation, however, that the bigger a particle's momentum, the smaller its wavelength.

A garbage truck moving at 30 miles per hour has a momentum given by the product of its mass, about 3,000 kilograms, and its velocity, about 10 meters per second. This works out to 30,000 kg.m/s. Planck's constant is 6.6×10^{-34} joule-seconds. Doing the multiplication gives for the wavelength $\lambda = 2.2 \times 10^{-38}$ meters, a number so incredibly tiny that such a wavelength would not have any physically observable consequences. It is because of this tiny wavelength that garbage trucks and other mesoscopic objects have sharp boundaries. Suppose, on the other hand, we took a proton moving at 30 miles per hour. The mass of a proton is 1.6×10^{-27} kilograms. Its momentum therefore would be 16×10^{-27} kg.m/s, and its wavelength would be 0.4×10^{-7} meters. While this is still very small, we shall see that it can produce measurable effects.

If we accept de Broglie's wave hypothesis, this still leaves us with two very difficult questions to answer. In exactly what sense is a particle a wave, and how do we measure its wavelike property, whatever that may be?

Demonstrating Wave Properties

There are specific classical techniques for demonstrating that something is a wave. For example, light is shown to be a wave by having it exhibit a phenomenon called "interference." When a single wave of light passes simultaneously through two narrow openings in an otherwise opaque material, it is broken up into two smaller "wavelets," each of which emerges from a single opening. The strange thing is that these wavelets do not subsequently behave as independent entities. Each retains a connection to the original wave and a memory of past behavior, evidenced by the fact that the separate wavelets continue to move in unison, a process called "being in phase." The peaks and valleys of one wavelet are precisely aligned with those of the other, as shown in figure 10.2.

The fact that the wavelets are in phase with each other causes them to form a so-called interference pattern when they overlap and strike a screen. When a peak of one wavelet overlaps a peak of another or when a valley of one overlaps a valley of another, the wavelets strengthen each other and create a bright spot on the screen. When peak overlaps valley, however, the wavelets cancel and the screen is dark. Figure 10.2 shows such an interference pattern as a curve of peaks and valleys representing the intensity of the wavelets where they strike the screen.

We say that interference is a property of waves and a reflection of their intrinsic "coherence" or wholeness. This is in contrast to the independence of particles that is a result of their discrete nature. If a stream of BBs was fired from a single gun at two slits, each BB would simply pass individually and independently through one slit or the other and retain no memory of any

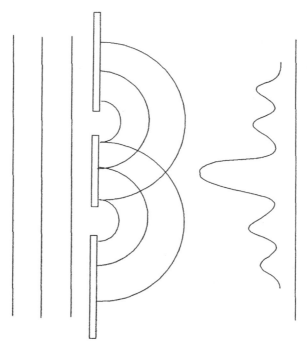

_____ *Figure 10.2* _____
Interference

connections it may have had with any other BB before it passed through. Even though the BBs were fired from a single gun at a regular rate, the slits would disrupt whatever organization the BBs previously had, and they would strike a wall behind the slits in a completely uncorrelated scatter.

If microscopic particles had wavelike properties, as de Broglie proposed, then it should be possible to make them exhibit some sort of interference by using a properly constructed set of slits. The size of the slits would have to be of the same order as the wavelength of the waves in order for the interference to become observable. BBs, for example, would have so large a momentum that their wavelengths would be too small to show a wavelike property with any imaginable slits. This was a good thing, since nobody had ever seen BBs exhibit interference patterns. Electrons, however, should be just right.

In order to produce clearly observable interference between waves, the slits through which the original wave passes must be very narrow, roughly comparable to the wavelength of the wave. In de Broglie's theory, the wavelength of an electron was extremely small, on the order of 10^{-10} meters, which is called an "angstrom" unit. There was no way that slits of this width could be fabricated by a machine or instrument.

As luck would have it, at just about this time Clinton Davisson and Lester Germer, two scientists working for the American Telephone and Telegraph Company, were studying the effects of electron beams fired at various metals. They were not really interested in basic theoretical research, but were trying to develop better filaments for vacuum tubes. At the time they began their research in 1925, they hadn't even heard of de Broglie's hypothesis and thought of electrons simply as small charged particles. They noticed, however, that when they fired electrons at a piece of nickel, the electrons bounced off in strange patterns that didn't fit in any scheme of electron behavior they knew about. Although the patterns were composed of individual "dots" like those you would expect from BBs, the dots occurred in regions that looked like the dark and light bands of an interference pattern.

While attending a lecture at Oxford University, Davisson heard about de Broglie's wave hypothesis and immediately realized that his experiment was a demonstration of the wavelike nature of matter that de Broglie had predicted. He and Germer thereupon returned to their laboratory and produced all the necessary results to verify de Broglie's hypothesis. What had in fact occurred in their experiment was that the regular arrangement of atoms in the piece of nickle played the role that slits did in the case of light. The spacing between the atomic planes was just the right size to produce the interference pattern.

Wave-Particle Duality

It appeared that de Broglie's hypothesis was true. Electrons were indeed waves. Or were they? The experiment disclosed a puzzling ambiguity. Although the electrons did indeed interfere with each other in some sense—otherwise there could be no interference pattern—they struck the screen in very definite spots. In short, they passed through the crystalline slits in a fashion that suggested wavelike coherence, but they subsequently struck a screen with all the definiteness and lumpiness of particles.

What were they? Many said that they were both wave and particle. It was suggested that they were a hybrid, called by some humorists a "wavicle," or waves at one time and particles at another. To some, they were one *or* the other; to others, they were one *and* the other; to others yet, they were neither. There was no clear consensus as to exactly what they were. The phenomenon, for lack of a better term, was called wave-particle duality.

Another puzzling aspect of particle behavior was seen in experiments performed many years after Davisson and Germer's. In these subsequent experiments, which used both electrons and neutrons, extremely sparse beams were used so that the particles were fired at the crystal literally one at a time. Even then an interference pattern was formed. The pattern could not be explained as the result of a conspiracy among neighboring electrons, nor was it a collective phenomenon, the result of there being an enormous number of electrons grouped together. One by one, in single file, widely separated in space and time, the electrons cleverly deposited themselves on the screen so as to gradually construct an interference pattern. Moreover, there seemed to be no way of predicting which electron would strike where. They struck the screen individually, with apparent randomness; yet out of this randomness the interference pattern inevitably emerged. One could almost imagine that each electron was equipped with a set of instructions telling it where it could land and where it could not. Some spots were permitted; others were forbidden.

Finding an explanation for the apparent randomness on an individual scale but coherence on the collective scale taxed the finest scientific minds. The explanation is still hotly disputed, so we shall examine it in detail.

Questioning an Electron: The Complementarity of Observations

Wave-particle duality is just one example of a property of microscopic reality that Niels Bohr termed "complementarity." In the mesoscopic world, we take it as a fundamental truth that the accuracy and completeness of a description can be improved by making more and more observations—that is, by asking more and more questions. When we are dealing with physical phenomena, these questions are posed in the form of experiments, whose results are the answers. Experimental results can be as straightforward as visual observations or as subtle as the readings of sensitive meters and other types of instruments. Whatever their form, however, we take it as unquestionable that we can gather all the results together at will, combine them, and construct a description of the phenomenon under investigation. The various results supplement each other in a classical principle of supplementarity.

Microscopic reality contradicts this belief in the validity of supplementarity. Observations of microscopic objects and the phenomena they engage in require that a choice be made of the particular set of questions you wish to ask and even when you want to ask them. Moreover, the number of questions permitted is quite limited. There may be many sets, but each has a small number of questions, and the answers from different sets can't be combined. Different answers may complement each other—provide alternative ways of thinking about phenomena—but they do not combine with each other for supplementation. This is the radical shift from classical physics. Each set must be regarded as providing an alternative mode of description. Each mode is valid and complete in itself (more on this completeness shortly), but it must be allowed to stand on its own.

The wave-particle duality of electrons is only one example of complementarity. It is a property not really of electrons but rather of the questions that we are permitted to ask them. If we wish to ask an electron, "Are you a wave?" we must first set up an experiment that is capable of asking (and answering) that question. A pair of slits in an opaque screen (or atomic planes in a crystal) is a device that will do just that. The device in effect defines the question we wish to ask, and performing the experiment provides the answer.

If we wish to ask the electron, "Are you a particle?" we must set up an experiment that can answer that question. A fluorescent screen that can produce a visible image when struck by a small, discrete object is the appropriate device. The experiment of Davisson and Germer really asked two questions in succession: "Are you a wave?" followed by "Are you a particle?" The electron answered "Yes" to both these questions, but did not do so simultaneously.

Why can't we combine these "Yes" answers and say that the electron is both a wave and a particle? The answer is simply "Complementarity." The two kinds of questions are not capable of being answered with a single precise statement. To get a single statement, we would need a single experimental apparatus that would simultaneously ask, "Are you a wave or a particle?" Such a doubly precise apparatus simply does not exist! The act of asking one question inevitably affects the set of physical conditions needed to ask the other. Einstein, you will recall, had to carefully define how length and time duration are measured in order to be consistent with his hypothesis that the laws of physics are the same in all inertial reference frames. Now we will have to carefully define how the properties of microscopic systems are determined in order to be consistent with our observations of those systems.

I cannot tell you why all this is so. Modern quantum mechanics simply accepts complementarity as part of the description of microscopic reality and tries to develop the simplest possible mathematical formalism to describe it and a philosophical framework in which to understand it. The concepts of wave and particle are inherently macroscopic and undoubtedly appeal to our macroscopic minds. Thus it is to be expected that at our first meeting with the microscopic world, we should think about it in macroscopic terms and categories.

Bohr's philosophy, which dominated quantum-mechanical thinking and came to be part of what is now called the "Copenhagen interpretation," incorporated a belief that the experimental apparatus became literally a part of the system under investigation and could not be separated from it. The nature of the apparatus in effect conditioned the system to respond to the question being asked. Bohr wouldn't even agree that there was an autonomous microscopic system holding answers to all conceivable questions and waiting for an experimenter to ask a particular set of them. He felt that the system didn't have answers until questions were asked—that is, until it interacted with the apparatus. When an answer finally appeared, it was the combined "system-apparatus" that was answering. To Bohr, quantum mechanics was a theory not of the behavior of microscopic systems but of how the microscopic and the mesoscopic worlds interact with each other, a theory not of microscopic reality but of what we can say about microscopic reality. There could be no theory of the microscopic world by itself in the sense that Newton's theory was a self-contained theory of the macroscopic world. Only a society of microscopic creatures living completely microscopic lives could develop a completely microscopic theory.

Just as Newton's conception of a common "now" shared by all observers presupposed a signal with infinite velocity to connect those observers with distant events, so would a microscopic system having properties independent of the mesoscopic world require an interaction with the mesoscopic world that had no affect on that system. According to Planck and Einstein, this was impossible. To Bohr, therefore, quantum theory was not simply the best theory; it was the only theory. Let me stress that this was Bohr's personal philosophy. It was not shared in its entirety by others in the field, particularly by Einstein or Schrödinger, as we shall shortly see.

Humans have a degree of complementarity in their actions, so the concept shouldn't be completely foreign to us. If we are going to ask for someone's hand in marriage, we try to influence the answer with soft lights, music, a fine wine, and so forth. Who is to say that the answer we get was there all the time and that the external conditions played no role in producing it? Think how the audience's interpretation of a play can be affected by scenery and stage setting. The surroundings of the action condition the audience to be receptive to certain nuances in the script, just as the experimental surroundings condition a system to present itself in a certain way. One of the challenges in psychoanalysis is that the interaction between subject and analyst is both strong and artificial, so that it does not necessarily reflect the subject's behavior in the "real" world. The analysis in effect is a particular experiment, and the results it elicits reflect the conditions of that experiment. These are examples of situations in human society that resemble those at the microscopic level. As we shall see, however, microscopic reality is stranger than any facet of mesoscopic reality that can be imagined.

An Example of Complementarity: Particle Spin

Complementarity plays such a fundamental role in the philosophical interpretation and mathematical description of microscopic phenomena that I would like to present the details of one particular example of it, a key part of the so-called EPR paradox, named after the initials of its three proponents, Albert Einstein, Boris Podolsky, and Nathan Rosen.

As I have already said, complementarity has to do with the fact that certain sets of physical quantities that separately describe a microscopic system cannot simultaneously be given values. Each description must be accepted as complete, standing on its own, even though the descriptions appear contradictory when considered simultaneously. The simplest possible system, like a single electron, is usually thought to be described by just two quantities, position and momentum. In macroscopic reality, these quantities can be simultaneously measured and defined. In microscopic reality, they cannot.

Elementary particles have another physical attribute that must be added to their position and momentum. Called spin, it has what superficially appears to be a mesoscopic analog called "intrinsic angular momentum." We discussed spin briefly in chapter 4 in relation to the helicity of the antineutrino.

Earth spins around its north-south axis once every 24 hours; this is called Earth's intrinsic angular velocity. Because Earth's core is molten and electrically charged, the angular velocity produces electrical currents, which in turn produce a magnetic field. This arrangement of currents to produce a magnetic field is called a magnetic moment. We can imagine that electrons also spin around an axis. They are charged too, so, like Earth's their spin is associated with a magnetic moment. Unfortunately, the analogy with Earth quickly breaks down.

To give spin a mathematical representation, it is treated as a vector, which can be pictured as a little arrow along the axis of rotation with a length proportional to the magnitude of the spin. The direction of the arrow is determined by the sense of the rotation, using a little trick called the "right-hand rule," which we noted in our discussion of antineutrinos. If you wrap the fingers of your right hand around the spinning object in the direction of its rotation, be it Earth or electron, your thumb points in the direction of the vector's arrowhead. Once the vector is constructed in this way, it is usually represented in terms of its "components" along three mutually perpendicular axes. You can think of the components as the shadows cast by the vector if a light shines down onto each separate axis.

In the macroscopic world, the vector of intrinsic angular momentum is completely measurable. You can measure either the length and direction of the vector on the one hand, or all three of its components on the other. In the microscopic world, however, complementarity rears its ugly head, and we discover that we can measure only the length of the vector and one of its components. Thus the length of the vector and its x component is one accept-able pair of measurable quantities, as are the length and y component or the length and z component. In classical macroscopic terms, any one of these mea-sured pairs would leave the exact status of the spin arrow undetermined. In microscopic terms, however, this *is* the exact status of the spin.

The situation can be shown with a simple experiment that uses what is called a "Stern-Gerlach" apparatus, developed by the physicists Otto Stern and Walter Gerlach in 1921. This apparatus is basically a channel between two magnets that are shaped so as to produce a nonuniform magnetic field between their poles. When a spinning particle passes through the channel, its magnetic mo-ment interacts with the magnetic field of the apparatus, and it is deflected ei-ther up or down. The amount of the deflection depends on the component of its spin (whose direction coincides with its magnetic moment) along the di-rection of the magnetic field. In other words, the orientation of the magnetic field is used to define a direction in an otherwise directionless space. The Stern-Gerlach apparatus then asks, "What is the component of your spin in this di-rection?" The spinning particle answers by moving up or down in an amount proportional to its component of spin along the direction of the field between the magnets' poles.

When this experiment is carried out using electrons, only two answers are ever obtained: the spin component is either up or down, always in the same amount, which turns out to be half a unit. (Stern and Gerlach actually did the experiment with silver atoms, for reasons that are irrelevant to this discussion; we will imagine they worked with electrons.) It doesn't matter how the mag-nets or electrons are oriented; when a beam of randomly oriented electrons is sent through the apparatus, members of the beam are deflected either up or down in the direction selected by the magnetic field, always by the same amount. Thinking again of a vector's component as the shadow it casts on an axis, the shadow of the spin is always half a unit, regardless of the angle one imagines it could or should be making with the axis.

This result tells us something very important about the nature of electron spin as opposed to its macroscopic analog, the intrinsic angular momentum of Earth. The components of electron spin in any given direction are quantized (they have only fixed, discrete values), just as Bohr assumed was true of the electron's orbital angular momentum. If, on the other hand, a designated com-ponent of Earth's angular momentum were measured, the value obtained would depend on the angle between the north-pointing vector and whatever appara-tus was used to make the measurement (something that measured Earth's magnetic field, for example). There would be as many different answers as there are different possible orientations of the apparatus.

To show that different components of electron spin are not only quantized but also represent complementary descriptions, we have to perform two Stern-Gerlach measurements in succession. Imagine then that we send a beam of electrons through a Stern-Gerlach apparatus oriented along what I will call the *x* axis. Emerging from the apparatus will be two beams, one containing all the electrons that answered "Up by a half unit" when asked the magnitude of their spins in the *x* direction and the other containing all those electrons that answered "Down by a half unit." We will now guide the "up" beam into a second Stern-Gerlach apparatus that is oriented along some entirely new axis, which I will call the *x'* axis. If electrons behaved like macroscopic creatures, the second measurement should produce a new amount of deflection, because the spin vector will have a different component along the *x'* axis from what it had along the *x* axis. The question "What is the magnitude of your spin relative to the *x'* axis?" should produce a different answer. But this is not what happens. The beam of "*x* up" electrons will again split into two equal beams, "*x'* up by a half unit" and "*x'* down by a half unit," as it passes through this second Stern-Gerlach apparatus, just as though the first passage had never occurred.

What does this tell us? Are the electrons in this beam spin-up in *x* by a half unit and spin-up or -down in *x'* by a half unit? Such a combination of properties cannot be satisfied by the components of a single vector. What the result tells us is that the two questions (and their answers) cannot be made simultaneously compatible. They are complementary. That is, both questions are reasonable by themselves and, when answered, provide a reasonable description of the electron at a particular instant. An electron whose spin is up by a half unit in the *x* direction is a reasonable electron; so is one whose spin is up by a half unit in the *x'* direction. An electron whose spin is simultaneously up by the same amount in both directions, however, is not found in nature, because there is no experimental apparatus capable of getting it to admit to having that pair of properties. Bohr would go so far as to say that it is meaningless to speak about an electron having that pair of properties simultaneously.

Suppose we now take that part of the original electron beam that had declared itself to be *x* down, which has not yet been fed into another Stern-Gerlach apparatus, and send it for a second time into an identical *x*-directed apparatus. This portion of the beam will once again be deflected down, indicating that asking the same question twice in succession will evoke the same answer. The entities of microscopic reality may have their vices, but they are not liars!

What, Then, Is the Nature of Microscopic Reality?

The characteristics of the microscopic world as revealed by the experiments and observations we have just been discussing can be summed up as follows:

1) The transfer of energy between radiation and matter occurs only in quantized units that depend on the frequency of the radiation.
2) When microscopic particles pass through appropriately sized slits, they produce interference patterns in a manner analogous to that of radiation.
3) These same particles, even as they form interference patterns in the aggregate, will individually strike a screen as discrete objects at unpredictable locations.
4) The electronic orbits in atoms are quantized in their energy and angular momenta.
5) The observations of certain sets of physical characteristics of a microscopic object, if taken together, produce an apparently contradictory description of that object. These observations cannot be combined to form a single, more complete description, but must be considered as forming alternative, complementary descriptions.

These five statements are either simple hypotheses to explain experimental results (1, 4, and 5) or statements of experimental results themselves (2 and 3). As they stand, it is difficult to quarrel with any of them. Of course, science will not be satisfied to simply let them stand; they must be incorporated within a larger theory that encompasses them all and that, in the best of all worlds, predicts yet new phenomena that can be experimentally verified. I will now present that all-encompassing theory, together with its successes and difficulties.

The Emergence of Quantum Mechanics

The five statements express I have just presented express the dominant features of the terrain of microscopic reality, and quantum mechanics is the theory that has become its map. As is the case with any theory of a reality beyond our ability to directly sense, quantum mechanics connects behavior at an "unsensible" level with observations we can make directly. We have already explored a similar connection between microscopic molecular behavior and mesoscopic thermodynamic behavior. That connection was provided by statistical mechanics.

If we cannot sense microscopic reality directly, then what do we sense? Quite simply, we sense only those things that are a part of the mesoscopic world. We sense the instruments and measuring devices that we ourselves construct. The only statements we can make with complete confidence about the nature of microscopic reality are statements about the behavior of the instruments. As already mentioned, Bohr saw this quite clearly, saying in effect that quantum mechanics was a theory not about microscopic phenomena but about the things we can say about those phenomena in mesoscopic terms.

One of the major arguments that arose over the interpretation of quantum mechanics relates to just how much we can say about a reality that presents itself to us only through instruments that belong to another reality. What happens in that reality when our instruments are turned off and we are not looking at them? Is that even a meaningful question? How do we describe a reality whose individual events seem to be random and unpredictable but for which repeated experiments inevitably bring forth patterns? Do we concentrate on the randomness or the orderliness? And if we consider nature to be essentially random, how can we explain the presumably ordered and deterministic nature of the lives we lead? Finally, how do we reconcile the macroscopic nature of the measuring devices with the microscopic nature of the system under observation? Can both the observer and the observed be incorporated within a single theory? These questions turn out to be exceedingly difficult to answer, and it is not clear that they have even now been answered in a definitive way.

Schrödinger's Wave Mechanics

The experimental results that have served to define microscopic reality, while admittedly strange and counterintuitive, are at least reasonably definite. They are there for all to see. The fascinating and difficult problems began to arise as scientists searched for a mathematical theory or model that would encompass those results and predict others. We shall see that the development of Einstein's quantum mechanical theory encountered many unforeseen difficulties.

The mathematical framework of quantum mechanics that is now accepted as definitive was developed by Erwin Schrödinger in 1925. It is called wave mechanics because the equation proposed by Schrödinger to explain the mesoscopic manifestations of the behavior of microscopic reality has a solution called a wave function, which has the properties of a wave. It fills space, moves with a velocity, and has a frequency, amplitude, and wavelength. In the equation, transcribed below for exhibition purposes only, the wave function is symbolized by the Greek letter psi (ψ). Schrödinger devised such an equation because he was explicitly looking for waves. The suggestion of de Broglie seemed plausible to him, and the subsequent experiment of Davisson and Germer seemed incontrovertible.

$$\frac{ih}{2\pi} \frac{\partial \psi}{\partial t} = -\frac{h^2}{8m\pi^2} \frac{\partial^2 \psi}{\partial x^2} + V\psi \,.$$

Schrödinger's equation represents for microscopic reality what Newton's second law represented for mesoscopic reality. It is a description of the motion of a massive microscopic particle under the action of an external force associated with a potential, V. Unfortunately, it was difficult to say exactly what physical aspect of nature Schrödinger's wave function referred to.

Newton's equation, $F = ma$, or, more precisely,

$$F = m \frac{d^2x}{dt^2},$$

which describes the motion of a macroscopic particle, left no doubt as to how its descriptive variables were to be interpreted. F was the net external force on a particle, m was the mass of the particle, and a was the acceleration of the particle. The acceleration was expressed in terms of the particle's position variables as $\frac{d^2x}{dt^2}$, so that solving the equation defined the particle's position as a function of time. That was called the "trajectory" or "orbit" of the particle, depending on whether it was a cannonball or a planet. Newton's equation as we've seen, reduces a particle to a set of three dependent position variables that change as a function of an independent variable, the time. The dependent variables in effect *are* the particle. The real physical entity has become a mathematical entity.

Schrödinger's equation, on the other hand, provides no such simple correspondence between a particle such as an electron and its coordinate variables. While the equation did contain the usual four variables, x, y, z, and t, which I have reduced to two, x and t, for simplicity, they were all treated as independent variables on which the equation's fundamental quantity depended. In other words, x did not depend on t. The two variables did not represent the position of a moving electron, but rather an arbitrary point in space and time. However, these two coordinates appeared on equal footing within yet another variable, a true dependent variable, the wave function. This fundamental quantity, the leading actor in the play, so to speak, was a function of the independent variables x and t, as indicated by their appearance within parentheses: $\psi(x,t)$. What was the physical interpretation of $\psi(x,t)$? If x is still to be thought of as a position of the electron at time t, then $\psi(x,t)$ must be something at that position. Schrödinger thought the wave function represented the electron directly. Like de Broglie, he believed that the electron really was a wave of matter. He turned out to be wrong.

In 1926, the physicist Max Born provided an interpretation for Schrödinger's wave function that persists to this day. Born proposed that the function determined the probability of finding an electron at a point in space at some particular time. The probability, however, was not the function itself but rather the square of the absolute value of the function:

$$P(x,t) = |\psi|^2$$

$P(x,t)$ represents the probability of finding the electron at position x at time t, the vertical lines bracketing the wave function, representing its "absolute value," are there because it is a complex number and has to be squared in a way that makes it real. Probabilities, after all, are real quantities, as is everything that is physically measurable. When the wave function, which is not measurable, is squared, we obtain a picture of microscopic reality as it presents itself to our macroscopic world. Much of the strangeness of the microscopic world is associated with the complex nature of ψ, which squaring does not eliminate but submerges. Because the squaring is required, we call ψ a probability amplitude.

By associating the wavelike nature of the function with an abstract quantity like probability rather than with the material body of an electron, Born promoted a theory in which the electron could still have the properties usually associated with a particle even as it engaged in behavior that would normally identify it as a wave.

This is a far cry from the no-nonsense interpretation we give Newton's equation, which tells precisely where a particle is at every instant of time on a trajectory, a precise mathematical curve. Any point in space through which the trajectory does not pass is a point at which the particle cannot appear. Schrödinger's wave function, however, has a value at every point in space at every instant of time, but it speaks to us only in probabilistic terms. Its square is the probability that the electron will be somewhere. A Newtonian particle is at precisely one place at one time. A Schrödinger particle can be anyplace, but depending on the values of the wave function, it is more likely to be found at some places than others. There have been attempts to think of the wave function as representing an object that simultaneously takes all possible Newtonian trajectories. While this is an interesting idea, it has never gained wide acceptance.

What is the relationship between Schrödinger's wave of probability and de Broglie's matter wave and what is the connection between the wave of probability and the interference pattern observed by Davisson and Germer? Schrödinger's wave function and de Broglie's wave are now considered to be one and the same, although de Broglie himself disputed that conclusion because he wanted his wave to be a real wave that carried a particle on its back. With Schrödinger's wave, the interference effects are produced not by any wavelike property of the electron itself but by the wavelike nature of its probability of existence.

Schrödinger's equation does not itself imply that one can think of an electron as being either a wave or a particle. It is not the corporeal matter of the electron interfering with itself that produces the interference pattern, but the parts of the probability amplitude passing through both slits. In mathematics of quantum mechanics, interference is a result of squaring the wave function to obtain probability. We also cannot say that the electron exhibits interference properties because of the wave function's mathematical properties. It is more to the point to say that the wave function is an appropriate description of the electron because with it we can reproduce the observed behavior of the electron. The wave function does not explain the electron; it merely describes the electron's behavior with surprising accuracy.

Let me give a simple example of how the wave function can be used to describe the behavior of electrons that pass through a system of two slits and strike a wall. The wave function has a piece that emanates from each slit and is therefore a sum of the two pieces. When the wave function is squared to obtain the probability that an electron will strike the wall at a given place, the two pieces multiply each other and create the interference pattern.

The wave function at the screen where the electrons produce their pattern, ψ, is a sum of two parts, one that emanates from slit 1, which I'll call ψ_1, and one that emanates from slit 2, which I'll call ψ_2. At the position of the screen, I can write,

$$\psi(\text{screen}) = \psi_1(\text{screen}) + \psi_2(\text{screen})$$

Each piece of the wave function has values over the entire screen, but the total wave function must be squared to determine the probability of finding the place where an electron strikes. The squaring process produces the result

$$|\psi|^2 = |\psi_1|^2 + |\psi_2|^2 + \psi_1^*\psi_2 + \psi_1\psi_2^*,$$

which of course is evaluated at the position of the screen. The superscript asterisks represent the complex conjugate of the wave function. The first two terms represent the probability that the electron has emerged from slits 1 or 2 and struck the wall. These terms will produce no interference effects. The last two terms, however, represent interference between the parts of the probability amplitude that emerge from different slits. This is the mathematical source of the interference effect. In it the Schrödinger equation provides us with a description of electron behavior that is consistent with the electron's actual behavior.

It is tempting to imagine that the two pieces of the wave function are two pieces of the electron, which is somehow splitting and passing through each slit, but this is simply not the case. Keep reminding yourself that the wave function is a description of an electron, not its physical essence. The Schrödinger equation describes the time development of a description. Moreover, the description is an indirect one both mathematically and conceptually, in that it must first be squared before it can be related to an aspect of reality. And even then, it becomes only a probability! This entire conception is far removed from the Newtonian description, in which the variables *are* the particle and the equation describes the motion of the particle directly.

Interference and Ignorance

When a wave function is written as a sum of pieces that represent mutually exclusive modes of behavior, the squaring of the wave function allows these modes to jointly affect the outcome of an experiment. This is consistent with experimental results. Classical systems can also exhibit mutually exclusive forms of behavior, but they present themselves in an either/or sense. Microscopic systems behave in a manner that suggests that the modes are simultaneously active. Somehow "the electron"—or at least some future pattern of its behavior—passes through slit 1 and slit 2.

If the slits are constructed so that it is possible to observe which one the electron actually passes through, then it is discovered that the electron passes through either slit, and no interference effects are seen. It is as though interference were a manifestation of observational ignorance. If the experimental apparatus is constructed so that it leaves us ignorant as to which of multiple possibilities actually occurs, then there is an interference effect that can be interpreted as implying that all possibilities occur simultaneously. But we cannot accuse the electron of behaving in so pathological a manner, because we do not observe the details of its behavior. If we do observe the intermediate details (i.e., through which slit the particle goes), then the electron behaves as would a classical particle.

This observation-dependent behavior is what Bohr referred to as complementarity. Different experimental arrangements yield different results, but the results cannot be combined to yield a complete picture of the system under observation. The results complement each other; they do not supplement each other. What ignorance hides cannot subsequently be revealed by a new experiment that offers to expose that ignorance. There have been many attempts to explain complementarity by some deeper analysis of microscopic behavior. One such attempt was initiated by the German physicist Werner Heisenberg.

The Uncertainty Principle

The fact that there are bumper stickers that say "Werner Heisenberg may have slept here" should be proof enough of the degree to which Heisenberg's uncertainty principle has been incorporated into the general culture. The significance of this principle and the central role it plays in the foundation of quantum physics cannot be overestimated. I will try to explain how Heisenberg developed it.

Heisenberg realized that because energy is transferred to matter in minimal quanta, the nature of measurement would have to be reexamined. It was a situation like Einstein's decision to reexamine the fundamental nature of length and time measurement when he realized that the speed of light was the same to all inertial observers. In the mesoscopic world, merely "looking" at an object does not affect its behavior. The act of looking is normally considered to be so benign that objects can be thought to behave when observed exactly as they do when unobserved. But is this really such a reasonable assumption? Certainly we know that humans do not behave in the same way when they are observed as they do when they are not observed. Why then do we expect microscopic particles to do so? Well, we say, humans behave differently under observation for psychological reasons, not for physical reasons. A robot wouldn't care a bit about being observed.

But think carefully about the act of observation and what it is supposed to accomplish. Some amount of energy must pass between object and observer: light or other signal must go from the object to the eye of the observer, carrying information about the object to the observer's brain. When that information registers on the brain, we say the observation has occurred. In the mesoscopic world, the amount of light needed to carry the information can be so miniscule that its presence is ignored by the observed object. In fact, it is usually assumed that the amount of light is zero, even though we all know that zero light would carry no image. Let's say then that the amount of light is "essentially" zero.

But is "essentially zero" good enough in the microscopic world? When I observe the Moon by the light of the Sun reflecting from its surface, it is not unreasonable to assume that the Moon's orbit is unaffected by the Sun's light. The energy of that light is essentially zero compared to the energy of the Moon. But if I observe an electron by shining light on it, the energy of the light could conceivably be comparable to the energy possessed by the electron. If that is the case, it is by no means clear that the electron's observed behavior is the same as its unobserved behavior. Light with energy comparable to the electron's could affect the electron appreciably, changing its velocity in a significant way. In fact, we can no longer speak of the electron's behavior without a modifying adjective, "observed" or "unobserved." Why can't we reduce the intensity of the light until its affect is not significant? Because hf is the least amount of energy transferred to matter by light of frequency f, and for any reasonable frequency, that amount is already comparable to the energy of an electron. The only lesser amount of energy available for that frequency would be an absolute zero, which would correspond to no observation at all. That would leave the electron unaffected, but it would leave the observer's eye and brain unaffected as well!

Why not use a lower frequency? you ask. Reducing *f* would certainly reduce *hf*, but unfortunately a principle of optics states that the minimum size of an object that can be clearly seen by light is approximately the wavelength of the light being used. (That is why a medical X ray shows much greater detail than an ultrasound image.) Thus to see a small object like an electron, one must use electromagnetic radiation of extremely small wavelength. But small wavelengths correspond to high frequencies. And because the electron is small, the radiation needed to "see" it must be high frequency. And because the energy transfer is at least *hf*, the electron will be significantly affected. The rules of this game are definitely not in our favor. We can't simply "imagine" what the electron is doing without actually shining light on it. Our imagination is not capable of encompassing the behavior of the microscopic world. We can discuss only what we actually measure.

If you want to measure the velocity of an electron, you must determine how much time elapses as it passes between two points a known distance apart. Unfortunately, in order to "know" the electron has passed the first point, you must make a measurement of its position. By whatever mechanism you have made the measurement, an irreducible amount of energy must pass to the electron. It is not necessary that the measurement be carried out with radiation. An electron's position could be measured by other means as well. Any measurement, by whatever method or device it is carried out, is a two-way process. It enables us to become aware *of* the electron and therefore conveys energy *to* the electron, energy that changes the velocity the electron had when it signaled its position. A simultaneous or immediately subsequent position measurement will therefore be made on an electron that has a different velocity from that it had previously.

The two-step velocity measurement does not provide information about an electron that is consistent with the electron's presumed position. You will of course obtain two numbers. The electron is "here," and its velocity is "so many meters per second." These numbers, however, do not have the same meaning as similar numbers for a mesoscopic object. If a mesoscopic object is repeatedly subjected to a simultaneous position and velocity measurement under exactly the same circumstances, each pair of measurements yields exactly the same results. When a microscopic particle is simultaneously subjected to a pair of measurements in the same repetitive manner, each pair of numbers is different. Moreover, the numbers differ in a completely random and unpredictable way. In short, a simultaneous position and velocity measurement of a microscopic particle is meaningless, meaning that it provides useless information.

Heisenberg summed this all up in what has come to be called the "uncertainty principle." In the simplest possible words, the principle says, "The more accurately you know the position of a microscopic particle, the less accurately you can know its velocity." Position and velocity are called conjugate variables, a name given to any pair of variables to which the uncertainty principle applies. Another pair is energy and time, and there are still others.

The uncertainty principle is a special and very limited case of Bohr's complementarity principle. Complementarity is a much more general and broader principle: it applies to entire descriptions of a system, while Heisenberg's principle applies only to specific pairs of variables, like position and momentum. But whereas complementarity was a philosophical vision, Heisenberg was able to demonstrate uncertainty in an experiment and draw a quantitative conclusion.

I have already noted that repeated measurements produce a random and unpredictable series of results. These results, however, differ by only a certain maximum amount. They do not vary over an infinite range of values because the radiation used to make an observation can impart only so much energy to the particle being observed. Let's call this spread in values the "uncertainty" in the measurement. If it is a position measurement that is uncertain, we symbolize its uncertainty by Δx. A velocity measurement's uncertainty is symbolized by Δp (p actually refers to "momentum," but for our purposes it's the same as velocity). Heisenberg's relationship is then written

$$(\Delta x)\,(\Delta p) \geq \frac{1}{2}(\frac{h}{2\pi})\,.$$

The product of the two uncertainties is at least Planck's constant divided by 4π. This is an amazingly simple relationship, but it is the reason for most of the paradoxical properties of the microscopic world as we interpret it by our observations. The uncertainty principle as originally postulated by Heisenberg is a statement about observations. Made by mesoscopic beings, the observations introduce uncertainties into the measured values of an electron's properties.

Do those properties have uncertainties even if they are not measured? That is a difficult question, to which several answers have been proposed. According to Bohr, the properties are there in the first place only because we measure them: the act of measurement produces uncertainties in their values. Bohr would advise you not to even think about a microscopic system's position and momentum. That would be thinking about the unthinkable. I will shortly discuss an argument between Einstein and Bohr in which Einstein tried to convince Bohr that it should be possible to think and talk about simultaneous properties of particles without actually measuring them. Einstein's contention has produced a long series of fascinating arguments that some physicists believe have still not been settled satisfactorily.

Probability and Uncertainty

A simple analogy might put matters into sharper perspective. A particular lecture hall in which I teach has two doors, one on either side. Students enter through either door A or door B and take their seats in the hall. Suppose you were a student arriving at the hall and you found only door A open. The probability of your entering through door A would be a certainty; mathematically stated, it would be a probability of one. Suppose you arrived on another day and found that only door B was open. Once again, there is no choice to be made. You enter through that door and take your seat, again with the probability of one.

On most days, however, you arrive and find that both doors are open. Now you have a choice to make before you enter the room and sit down. The option may affect your mental decision-making processes and cause you to pause momentarily, but you will definitely choose one option or the other. When I arrive to give my lecture and see you sitting in your seat, I know that you have either entered through door A and sat down or entered through door B and sat down. Even if I haven't actually observed you walking through one of the doors, I know for a fact that you must have done so. Your body, being mesoscopic in nature, must have taken a Newtonian trajectory that moved you along a definite path through one of the doors to your seat. The fact that both doors are open may cause you to experience a degree of mental indecision, but it does not cause your possible physical paths to interfere with each other. There are no seats in the hall that you cannot be found at because the two sets of possible paths cancel each other at those positions.

Consider now an electron's behavior. Instead of doors and seats, the electron is faced with two extremely narrow and closely spaced slits and a fluorescent screen just beyond them. I ask you to accept my assurances (the experiment has actually been done) that if either slit were blocked, the electron would pass through the other slit, and there would be no interference pattern on the screen. With both slits open, we have the mathematical ability (through Schrödinger's equation) to determine the wave function of the electron between the slits and the screen. I have already discussed this two-piece wave function, which looks very much like the sum of the individual wave functions that would be obtained if each slit were alternately opened. Nevertheless, when this wave function is squared to obtain the probability that the electron will strike the screen at some particular place, the resulting function has the interference pattern.

There may be significant effects on your decision-making process due to the two options available to you, but ultimately your body takes one Newtonian trajectory that leads through either door A or door B along a particular path that passes through a particular door and selects a particular seat. This is not the case for the electron. We cannot say that it passes through one particular slit and strikes the screen. We could make such an assertion only if we placed at each slit a detector capable of indicating the particular slit through which the electron had passed. Including such detectors in the experimental apparatus, however, would change the entire experiment. Schrödinger's equation would be different. The nature of the questions we could ask and the answers we could get would change.

Without an actual observation of the particular slit through which each electron passes, we *cannot* say that the pattern on the screen is formed by the arrival of electrons, each of which has taken a definite path through one slit or the other. If we could say that, the interference pattern would not be there. Without detectors, we can say only that the pattern is composed of electrons that have passed through the two slits and struck the screen. This is not the same as saying it is composed of electrons that have a possibility of passing through slit A and striking the screen or passing through slit B and striking the screen.

An experiment is conceivable that would allow you to deduce which slit the electron had passed through to within a certain degree of accuracy. You wouldn't be able to say with certainty that it went through one slit or the other, but neither would you be completely clueless. With that experiment, the interference pattern would be partially destroyed. It is really quite amazing how the nature of complementarity pervades the entire structure of the microscopic world.

The Reality of the Wave Function

The theory of quantum mechanics is intended to provide both a mathematical description of microscopic reality and the necessary instructions for making predictions about its behavior. If a description were all that was necessary, it could be given by a list of the ways microscopic systems behave under various experimental arrangements. Such a list might read something like, "When passing through two slits, an electron produces an interference pattern; when striking a fluorescent screen, an electron makes a single particle-like spot"; and so forth. The list would be a rudimentary map, a collection of brief descriptions of all the terrain that had actually been visited by experimental physicists. But such a map is both unsatisfying and insufficient. It says nothing about regions that have not been investigated; it provides no insights and it permits no predictions about experiments yet to be performed. The triumph of Schrödinger's equation was the fact that it not only described what had already been observed but provided a consistent explanation of those observations and made predictions about the results that could be expected from other experiments not yet performed.

Schrödinger's equation, however, is couched in terms of what is perhaps the most mysterious mathematical entity in all of theoretical physics, the wave function. Even today, many years after it was postulated by Schrödinger, the meaning of the wave function is hotly debated. According to most physicists, it should be interpreted as a probability of different modes of behavior of a microscopic system or of different outcomes of a given experiment on the system. Heisenberg spoke of it as describing a system's "tendency" to behave in a certain way or "potential" for certain forms of behavior. The actuality of a system's behavior is completely beyond the province of quantum-mechanical description; tendencies, however, can at least be verified statistically by observing the behavior of many identical systems. Indeed, this statistical point of view is the interpretation given to the wave function by most physicists when they use it to make quantum-mechanical calculations.

Yet even with the wave function's successful use under the assumption that it provides a description of the behavior of large numbers of identical systems, many unresolved conceptual questions remain. Two of the most troublesome are Must we be resigned never to explain an actuality associated with an individual system? and How shall we explain the behavior of mesoscopic and macroscopic objects, humans included, which are individual systems that behave in a definite way? One of the things I find most disturbing is the degree to which the wave function has become a surrogate for the system itself—so much so, that problems encountered in trying to understand the wave function are often mistaken for problems with nature itself. One would think intelligent people would not mistake the description for the object, even as they would not shoot the messenger that brings them bad news. I guess it is partially a result of the fact that because we have no way of directly sensing the microscopic world, we have let the wave function become its proxy. All this being granted, however, don't let me mislead you. Any time the description of a new reality is filled with paradoxes, you can bet that the reality itself is not innocent. In the following sections I will discuss some of the difficulties that have been encountered by attempts to find a consistent and sensible interpretation of the wave function, and I will finally comment on the implications this situation has for microscopic reality itself.

The "Collapse" of the Wave Function

Every interpretation of the wave function seems to bring with it its own collection of unwanted paradoxes. If you subscribe to the notion that the wave function is applicable only to the statistical description of an ensemble of systems, you are giving up the ability to make any sort of meaningful statements about an individual system. Recent experiments have been sensitive enough to produce measurements of individual atoms. How do you answer the questions of the physicists who did these experiments if you don't believe the wave function applies to their system? But if you believe that the wave function does describe an individual system, like a single atom or electron, then a paradoxical type of behavior called "collapse of the wave function" can occur.

Suppose, for example, a single electron is confined to some sort of box or closed room. If we assume that the electron can be found anywhere in its container with equal probability, its wave function will be a constant. If an observation is made that locates the position of the electron, immediately following that observation, the electron's wave function must be entirely different; now it describes an electron whose position is known precisely and with certainty. This being the case, the previous constant-valued wave function becomes zero at all positions of the box where we now know the electron was not found. This instantaneous change of the wave function, called its "collapse," appears to occur when a measurement creates a new situation for the system under consideration.

The formalism of quantum mechanics makes a strict distinction between changes in the wave function that occur as a result of the processes described by the Schrödinger equation and changes that occur as a result of measurements. The former changes describe the behavior of the system between measurements. They are smooth and deterministic. The latter changes are abrupt "collapses" apparently not encompassed by the Schrödinger equation; moreover, experiments show that they are unpredictable by their very nature.

The existence of two types of wave-function changes is not the only problem, however. Another problem results from the fact that sudden collapses appear to violate basic principles of relativity. In effect, any instantaneous change of the wave function over a finite region of space would require the transmission of a signal at a speed faster than that of light, telling the wave function that a measurement was just made and it had better collapse.

Heisenberg realized that such collapses were not acceptable and would require a reinterpretation of the wave function. He therefore proposed that the wave function be viewed as representing "our knowledge" of the system rather than the system itself. Thus the square of the wave function is our knowledge of where the system is likely to exist rather than the actual probability of its existence there. When a measurement is made, its results instantaneously change our knowledge of the system's behavior, and this knowledge might extend over vast regions of space. Since the knowledge is located in a small region of our brains, however, the signals that must move around need not travel at excessive speeds.

Clearly, by shifting the subject matter of the wave function from the system to our knowledge of it, we appear to avoid a serious problem with relativity. However, we shall see that it does not avoid all problems.

The Paradox of Schrödinger's Cat

Another set of collapse-related paradoxes arises when we try to include a mesoscopic measuring apparatus within the quantum mechanical formalism that describes the system being measured. To Bohr, the measuring devices were necessarily mesoscopic and therefore belonged to that reality. Their behavior would be described by Newton's laws. Yet ultimately, if quantum mechanics is to be considered as an all-inclusive theory, it should be able to describe everything. There should not be a need for one set of theories for the system and another for the apparatus. After all, every mesoscopic object, including us, is made up of the atoms and molecules that quantum mechanics supposedly describes. In principle at least, we should be describable as aggregates of atoms through some sort of quantum-mechanical wave function.

Many attempts have been made to incorporate both measuring devices and microscopic systems within a single wave function. A complete wave function would have to contain all the necessary descriptive variables for microscopic and mesoscopic worlds. The problem with such an approach is that the wave-function formalism automatically renders its variables as probabilistic. The definiteness that we associate with the mesoscopic world appears to be lost. The attributes of a mesoscopic device described by a microscopic wave function lose their Newtonian certainty.

Schrödinger, in a moment of black humor, tried to underscore this difficulty with an example that has come to be called "Schrödinger's cat paradox." He envisioned a diabolical device, a windowless box with a cat inside and a mechanism that could kill the cat by exposing it to a poisonous gas. The gas would be released when the mechanism detected the decay of a single radioactive atom within the box. The nature of the atom was such that it had an approximately 50% chance of decaying within a certain period of time. At the end of this time, an observer would open a door in the box, look in, and determine whether the cat was dead or alive.

The action of the diabolical device was intrinsically microscopic, because a single radioactive atom must be described quantum-mechanically and the time of its decay cannot be predicted with mesoscopic certainty. If the combined cat-atom system were to be described by a single Schrödinger wave function, that wave function would have to be the sum of two parts. One part would describe a live cat and an undecayed atom, the other part a dead cat and a decayed atom.

The paradoxical nature of this sort of description was that the cat's existence was now put on the same probabilistic footing as the decay of the atom. From the wave function alone, we cannot say that the cat is either dead or alive, only that it exists in a kind of probabilistic limbo until some sort of observation is made, such as peeking into the box. The probabilistic existence of an unobserved system may be acceptable for atoms—after all, nobody has ever seen an atom—but cats are a different matter altogether. Being reasonable people, we know that even if we didn't look into the box, the cat would be either dead or alive.

Schrödinger's cat paradox involves several aspects of quantum-mechanical description that have caused great minds to differ. First of all, the assumption that the wave function is a legitimate means of describing a single system is not a paradox for those physicists who would not grant the wave function meaning for a single system. They would claim that the wave function applies only to an ensemble of many cat-in-the-box devices. The two parts of the wave function would not describe a single cat in a dead-or-alive state, but rather the results of doing the experiment many times, in which case some cats would die while others would live. This sounds quite reasonable, but it requires abandoning the hope of ever describing individual systems, which would render quantum mechanics a much less useful approach to understanding microscopic reality.

Heisenberg would approach the paradox by asserting that the wave function does legitimately describe a single cat, but only the tendency of the cat to be either dead or alive. The actual condition of the cat can be ascertained only by someone looking into the box. This observation would modify that person's knowledge, and a new wave function would then be called for. Until an observation is made, however, quantum mechanics allows us to say nothing definite other than to talk about tendencies.

Bohr would simply deny that a cat can be combined with an atom in a single wave function as though it too were a microscopic object. The cat, being a mesoscopic object, must be accorded all rights and privileges due such an object, most importantly a description by Newtonian physics (or classical biology). Thus the cat is most certainly alive until the atom decays, and the atom, which is legitimately described by a wave function, leads the probabilistic existence expected of all atoms.

Decoherence: What It Means to be Macroscopic

A group of physicists has recently shown that macroscopic objects like cats can be treated quantum-mechanically. However, their wave functions are an extraordinarily complex combination of the wave functions of each of their individual atoms and the interactions among the atoms. The complexity of a cat and its wave function produces a phenomenon called "decoherence," meaning (in simplest terms) that macroscopically different states of the cat (e.g., a live cat and a dead cat) cannot interfere with each other in the quantum sense of the term. In short, the sheer complexity of macroscopic objects prevents them from exhibiting the interference effects that are the hallmark of life in microscopic reality. That being the case, when the cat and a radioactive atom get together, there is no wave-function superposition of live cat and dead cat, which would produce interference effects when squared, there is only the classical pair of alternatives appropriate to macroscopic entities: a live cat or a dead cat. It is unnecessary to peek into the box in order to create reality and collapse the wave function. Even when unobserved, the complex macroscopic cat has already untangled itself from the quantum world and become a truly macroscopic object.

This is an important result, because we know that a dead cat can be autopsied to determine the exact time of death. If it is our observation that killed the cat, then all autopsies would have to show that the cat died immediately upon our opening the box. The mesoscopic and macroscopic worlds simply do not behave in this way.

Is Quantum Mechanics a Complete Description of Reality?

Albert Einstein, although one of the founders of quantum mechanics, nevertheless believed it was only a provisional theory, providing a complete description but a superficial explanation of microscopic reality. In particular, Einstein rejected Bohr's contention that the best we can do in describing microscopic reality is to present the results of experiments. To Einstein, a complete description would also present the external reality as it exists independent of experiment. Einstein believed there was a real world that behaved in accordance with certain laws even in the absence of our observations of it. We should be able to describe that behavior and deduce those laws, not limiting ourselves to appearances and the results of experimental observation. Just as Boltzmann's statistical mechanics provided a deeper level of understanding than did thermodynamics, so would a more complete version of quantum mechanics provide a deeper level of understanding of the microscopic world. Thus to Einstein, the apparently random and unpredictable results of measurements made on microscopic systems had a completely deterministic explanation. Randomness on the mesoscopic level somehow reflected precision on the microscopic level.

Einstein envisioned "hidden variables" not contained within the quantum-mechanical description of nature, which, if their values were known, would provide the sought-after complete description. He was willing to accept a probabilistic interpretation of the wave function if the probabilities referred to an ensemble of identical systems, each of whose variables possessed definite values. Like Boltzmann's description of matter, Einstein's hidden variables would be analogous to the enormous number of simultaneously definite positions and velocities of each molecule. The thermodynamic description did not recognize Boltzmann's variables and so provided a less precise explanation of thermal phenomena. Once the complete statistical description was known, the thermodynamic description could be obtained from it by dealing with macrostates rather than microstates. This in effect was a voluntary giving-up of a description that contained a virtually uncountable number of variables in favor of one that contained a few. What was important was the fact that the microstate description was there should one choose to use it. A complete theory would predict microscopic phenomena with the same precision with which Newton's law predicted mesoscopic phenomena. Einstein once said that "God does not play dice," meaning that the world we live in does not depend on chance at its core. He was a determinist at heart.

Opposing Einstein's belief in the imperfections of quantum theory was another of its founders, Niels Bohr, who believed that quantum theory provided the best description of nature possible. Every level of description required the use of mesoscopic measuring devices that by their very nature affected the descriptions they provided. Seldom had conceptual battle lines been so clearly drawn. Einstein decided to challenge Bohr by conjuring up an experiment whose outcome would invalidate Bohr's assertions. This was not an experiment that Einstein actually did or could do, because the technology at that time was not good enough. Rather, it was an experiment whose outcome seemed so clear and obvious that Einstein carried it out in his mind as though he had actually performed it. Such an experiment is called a "Gedanken experiment," from the German word for "thought." Einstein presented this experiment along with his entire chain of reasoning in a paper he wrote together with two friends, Boris Podolsky and Nathan Rosen. The experiment and its presumed results, all in Einstein's mind, are now called the "Einstein, Podolsky, Rosen paradox," or EPR for short. It has assumed such a central role in the challenge to the fundamental meaning of quantum mechanics that I will discuss it at some length.

The EPR Paradox

If quantum theory was complete, as Bohr insisted it was, it should be able to describe all aspects of physical reality. To Einstein, physical reality at the very least consisted of all those quantities that could in principle be "predicted without in any way disturbing the system"—in short, those attributes of nature that exist independent of an observer, the oasis without the mirages. Accepting this definition of physical reality, Bohr would assert that position and momentum, considered as a pair of simultaneously defined quantities, have no physical reality, because there was no way of predicting the value of one without disturbing the system as you measured the other. Given that fact, there would be no need for quantum theory to provide a description that included simultaneous values of position and momentum. Einstein thought that he could find an aspect of physical reality that quantum mechanics didn't describe. Thus the battle lines were sharply drawn.

With his colleagues Podolsky and Rosen, Einstein suggested an experiment whose outcome should demonstrate the incompleteness of quantum mechanics. The EPR experiment was quite simple conceptually, even though, as noted above, the technology for performing it was unavailable at the time. It was finally performed in 1972, more than fifteen years after Einstein's death, and has since been repeated several times in different forms and with more precise results.

The EPR experiment is predicated on the fact that the values of certain physical quantities associated with a system are conserved (they do not change) during the evolution of that system. Consider, for example, a molecule like hydrogen, which is composed of two atoms. One of the conserved physical quantities associated with such a molecule is the sum of the angular momenta of the nuclei of the two atoms. Angular momentum can be visualized as the spinning motion that each nucleus has (although "spinning" is really a mesoscopic phenomenon, and it is highly unlikely that microscopic particles spin in the way that a top or Earth does). The nature of this imagined motion can be described by a vector, meaning it can be associated with a magnitude and a "direction." Continuing the mesoscopic analogy, the direction of the angular momentum or spin vector would be the direction of the axis around which the rotation is occurring. If you imagined a set of three perpendicular axes and maintained a classical picture of the vector, it should generally have components along each of them.

In terms of quantum mechanics, however, the three components of spin are conjugate variables, and thus, according to Heisenberg's uncertainty principle, measuring one of them makes it impossible to simultaneously measure another with precision. No pair of the three can be regarded as simultaneous elements of reality. Any single component of an atom's spin in a given direction, however, can be measured by passing the atom through a magnetic field oriented in that direction, which will deflect it by an amount related to the spin component.

It is possible to conceive of a hydrogen molecule in which each atom's nucleus is spinning in an opposite direction from the other, so that their total angular momentum is zero. If the molecule were split apart in the proper way— this is the part of the experiment that had to be carried out in Einstein's mind— the sum of the individual angular momenta would remain zero, even if the two atoms flew into widely separated parts of the laboratory. This "conservation of angular momentum" is a result of the fact that nothing happens to either of the atoms during the splitting or subsequent flight that could change the value of the angular momentum they had when they were together as a molecule.

Einstein imagined that the splitting apart and fight were somehow accomplished. When one of the atoms, call it atom A, enters its target portion of the laboratory, an experimenter there measures a particular component of its angular momentum, the relationship between the atom's axis of spin and some arbitrarily chosen axis in the laboratory. The act of measuring disturbs the atom, making a simultaneous precise measurement of some other component of its angular momentum impossible, but the instant the value of atom A's angular momentum in the chosen direction is known, the value of atom B's angular momentum in that direction is also known, because the sum of the two must be zero. In other words, an experiment carried out on atom A, together with knowledge of the relationship between A and B, allows a precise prediction to be made of a property possessed by atom B. At the instant that the experimenter deduces the value of B's angular momentum, the measurement made on atom A could in no way have physically affected atom B, given the separation of the two and the fact that no influence can move faster that the speed of light. So far there is nothing paradoxical about this entire process, but in a moment there will be.

Suppose the experimenter dealing with atom A had decided to measure the component of its spin along a different axis, conjugate to the one actually chosen. An equally precise piece of knowledge about atom B would have been obtained, again without disturbing atom B. Each of these separate measurements on atom A would affect the precision of simultaneous measurements on conjugate variables of atom A, but in principle they would enable the prediction of precise values for a pair of conjugate variables of atom B. Of course it could be argued that only one of the experiments on atom A could actually be carried out; the second measurement is a hypothetical one.

But it is the possibility of carrying out either experiment that produces the paradox. Since atom B could not possibly know in advance which experiment was to be carried out on atom A, and since the experimental act itself could not affect atom B, the values of the conjugate pair of variables that would be revealed by the two hypothetical experiments on A must already be precisely defined for B. This pair of variables must therefore be elements of physical reality, because they can be known for atom B without disturbing it. Since quantum mechanics forbids simultaneous precise knowledge of pairs of conjugate variables, Einstein concluded that quantum mechanics was incomplete.

Perhaps an analogy will help. As I rush out of my house one morning, I pick up a pair of gloves that I left on the table next to the front door and stuff them into a pocket of my coat. Unfortunately, in my haste I pick up only one of the gloves and leave the other behind. A bit later, I reach into my pocket for the gloves and feel only one. I immediately realize what I have done. Even before I take out the single glove and identify which hand it belongs to, I know that the one I left behind will belong to the other hand. This knowledge results from the fact that a pair of gloves is always "correlated"; there is one for each hand. I won't know which glove I left behind until I look at the one in my pocket. If I call that observation an "experiment" carried out on the glove in my pocket, then I certainly obtain information about the other glove without affecting it in any physical way. The separation of the pair of gloves produced by my hasty grabbing preserves the correlation between them, although I must perform one experimental observation to discover which glove was left behind.

If the pair of gloves behaved like a pair of hydrogen atoms, however, Bohr would assert that neither the right-handed nor the left-handed glove was actually in my pocket until I pulled it out and looked at it. All that could be said prior to the observation would be that my pocket contained something that had a 50% chance of being a right-handed glove and a 50% chance of being a left-handed glove when I pulled it out. Only when I pulled it out could I ascertain its identity and know which glove was left behind. Einstein, of course, would claim that the glove in my pocket was either left-handed or right-handed from the moment I picked it up. My observation of it served simply to affirm what it was all along. Now in the case of gloves, it seems reasonable that Einstein's argument wins out, since there is no way the glove left at home could instantaneously accommodate itself to the discovery I make of the glove in my pocket. This was Einstein's reasoning about the EPR experiment. However, it is not at all clear that this is really the case for hydrogen atoms, which are governed by the uncertainty principle.

If you subscribed to Einstein's reasoning, the EPR experiment posed a classic dilemma. Either quantum mechanics was incomplete, in the sense that conjugate physical quantities really were simultaneously definable but quantum mechanics could not define them, or else influences from an observation could fly faster than the speed of light from one part of a system to another widely separated part.

Bohr responded to this dilemma with an argument that did not then and does not now appear overly satisfying. Einstein's definition of "physical reality," according to Bohr, was incorrect. A correct definition refers not to the isolated external system but to the information we can obtain about it. In this sense, physical reality is conditioned as much by the particular experiment we decide to do as it is by the system on which we experiment. To be able to say that what we do does not disturb a system requires more than just not disturbing it physically by the act of measurement; it means that we must not disturb it "informationally," in the sense that what we do cannot be allowed to affect what we can learn about the system. Our choice of which measurement we will do in one room does disturb the informational aspects of the system because it makes available a new set of experimental results. Thus, Bohr maintained that Einstein had not defined an element of physical reality not described by quantum mechanics.

Bell's Theorem

In 1964, the physicist John Bell derived a theorem that ultimately laid to rest the EPR challenge, revealing an extraordinarily strange aspect of the quantum-mechanical description of nature. Notice that I did not say a strange aspect of nature, but rather of our description of nature. The difficulty of separating description from the thing being described is at the heart of this problem.

Bell assumed that all of EPR's major premises are correct. First, Bell agreed that widely separated atoms cannot communicate with each other in any way that violates relativity. This premise is commonly called "locality," meaning that physical operations on a system affect that system only within a small region of space during any short period of time. Bell assumed that the nature of the system in one part of the laboratory cannot be affected by choices made in another part. If one experimenter decides to measure the y component of spin instead of the x component, another part of the system cannot immediately reorient its spin to compensate for this sudden change of design.

Second, Bell concurred with Einstein's philosophical belief that the atom already possessed those values of its physical properties the experiment was being asked to reveal. This belief is a form of determinism, because it assumes that the behavior of natural systems is predetermined: at every instant of time, a system "knows" how it is about to behave in the next instant. In clear opposition was Bohr's belief that the values of physical properties are spontaneously and randomly created by the experimental act itself.

Third, Bell accepted Einstein's assertation that, to be complete, quantum-mechanical description would have to contain all the variables necessary to specify the values of the physical quantities about to be revealed. Bell therefore arbitrarily inserted into the wave function a set of such variables, which Schrödinger's equation of course did not contain. By augmenting the wave function in this way, Bell created what is commonly called a "hidden-variables" theory. All this would have made Einstein happy, because it represented everything he wanted his description to contain.

Finally, Bell granted Einstein the premise that an experimenter could choose to perform any measurement and that the measurement would provide results consistent with the physical values the system was already assumed to have. This seems so obvious and reasonable an assumption that it is seldom mentioned; yet is is an important part of the EPR scenario. It has been given the somewhat intimidating name of "contrafactual definiteness," meaning that what in fact does not occur would have the predicted results if it were to occur. Bohr's quantum mechanics has nothing to say about quantities that are not measured. Bohr would say that any measurement will create some value of the quantity being measured, but that quantity is indefinite until the measurement is performed. The following analogy might help to distinguish these points of view. Suppose you miss your plane and it crashes. You have two possible responses. 1) How fortunate I am; had I caught the plane, I would surely have been killed. 2) I am neither fortunate nor unfortunate; had I caught the plane, the entire set of conditions would have been different, and it is impossible to know whether the plane would have crashed or not. The first response is consistent with contrafactual definiteness, the second with Bohr's philosophy.

With the four premises of locality, determinism, hidden variables, and contrafactual definiteness accepted as true, Bell proved the following remarkable theorem: a local, deterministic hidden-variables theory cannot reproduce the statistical predictions of quantum mechanics. This statement is so brief that its significance might not be evident. The statistical predictions of quantum mechanics, after all, are assumed to be correct, Not even Einstein doubted that; he simply felt that the predictions did not go far enough. He wanted to transcend the statistical and describe the individual. But Bell proved that the version of quantum mechanics Einstein wanted would not even be correct in the statistical sense. Boltzmann's statistical description, for example, predicted everything that thermodynamics did and more. A local, deterministic hidden-variables theory, on the other hand, would not be consistent with the theory it was trying to improve on. In short, if Einstein were to make quantum mechanics complete, he would also make it incorrect—unless quantum mechanics already *is* incorrect, in which case Einsteins's version might be an improvement!

The Proof of the Pudding

Unlike the Gedanken experiment in the EPR paper, Bell's theorem suggested certain real experiments that could be performed to determine whether Einstein's or Bohr's version of quantum theory was in accord with reality. The first such experiment was performed in 1972 by S. J. Freedman and J. F. Clauser. It was not a faithful replica of Einstein's Gedanken experiment, but it could be used to verify Bell's theorem. To make a long story short, the results supported the idea that a local hidden-variables theory is not consistent with nature but the statistical predictions of quantum theory are. Bohr was finally proved right, Einstein wrong.

What are we to conclude from all this? It is sometimes asserted that Bell's theorem demonstrates that relativity must be violated at the level of microscopic reality. This is not true. The theorem proves only that the statistical predictions of quantum mechanics are inconsistent with a local deterministic theory. The kind of locality assumed by EPR and Bell just forbids the experimental observations in one region from depending on the choices that could be made by an experimenter working in another region. The visible behavior of something in one room immediately and directly depends not on what I actually do in another room, but on what I could do in another room. That is an even more mysterious effect, but it could not be used to transmit messages at speeds greater than light, so it does not violate relativity, although it does violate common sense.

All of the problems pointed out by Bell arise because we are not satisfied to accept a statistical description of reality. It is only because Einstein wanted to think about the real microscopic world in the same way we think about the mesoscopic and macroscopic worlds that the problem arose. He wanted to be able to think about a single system apart from the experiment being done on it. But if you insist on thinking like that, you must think nonlocally, and that means that what someone else *is* doing depends on what you *could* do. Thus the quantum-mechanical description of the microscopic world has implications for the unthinkable. Not only can't physical reality be separated from the experiments we do on it, but we can't even think about that possibility unless we are willing to accept the fact that choices we can make in one room will affect behavior in another.

Summary and Conclusions

It is essential to remember that the electron—or for that matter any microscopic entity described by Schrödinger's equation—has no objective physical qualities other than those produced by a measurement. Recall how Einstein stressed the fact that space and time have no qualities other than those bestowed by clocks and rulers. In exactly the same way, the entities of microscopic reality have no abstract properties. When an electron strikes a screen, a "measurement" of its position is made. The electron is found to be at the point where it strikes. That measurement establishes its position and also the more basic fact that it is an electron. We know that because it has just transferred a small amount of energy to the screen, which shows up as a small fluorescent spot. What, if anything, could we have said about the electron before it struck the screen? Nothing at all, unless we happen to have established certain qualities at a previous measurement that we have reason to believe have not changed. The things about an electron that we call its "properties" are established by the nature of the measurements we perform. An electron has "a position" because a fluorescent screen establishes position. Niels Bohr, with his idea of complementarity, was the first person to stress the importance of the act of measurement and the experimental apparatus itself in establishing the properties of a microscopic entity.

With measurements, we make the electron conform to our notions of what a mesoscopic entity is. Mesoscopic objects have position, so we demand that the electron have position, erecting obstacles it will bump into. The electron communicates to us through our measuring devices, but those devices force the electron to speak to us in a mesoscopic language. Therefore we hear only words that our devices form, which may not be all those the electron has to convey.

The major limitation quantum theory has placed on our ability to know is that it has completely modified the meaning of the world out there, which is precisely the thing we wish to know about. It is not the quality of knowledge that has changed, but the nature of the entity to which that knowledge relates. In microscopic reality we perform experiments to learn not about an electron but about experiments performed on an electron.

If we insist on retaining that tenuous grip on sanity and common sense that allows us to imagine there really is an electron out there that we are studying—not that we are studying ourselves studying an electron—then we run into Bell's theorem, which forces us to think of the electron as an object that somehow correlates choices we could make as to how we will think about it. In short, we are either studying an entity that is inextricably tied up with our observational devices, or else we are studying something that affects our thoughts themselves.

Finally, we must come to grips with knowledge that is created rather than revealed and fundamentally uncertain rather than precise. In classical physics we had become accustomed to saying, "I do not yet know this!" In quantum physics we must become accustomed to saying, "I cannot know this!"

The old National League umpire Bill Klem told a story about a batter who persisted in arguing with him over the way he was calling balls and strikes. "Any idiot could see that last pitch was a ball!" the batter screamed. "Listen," Klem responded calmly. "Until I called it, it wasn't anything; it became a strike when I called it a strike." I doubt if Klem was a student of Bohr's philosophy, but that remark summed it up beautifully.

Eleven

The Chaotic World

Is our macroscopic world governed entirely by deterministic laws, meaning that the present is strictly determined by the past, or is it subject to the vicissitudes of chance? This is not an idle speculation. If chance and randomness do indeed play a significant role in the world, then we must reconcile ourselves to the fact that our ability to understand natural phenomena will be fundamentally limited. If, on the other hand, meso/macroscopic phenomena are determined and predictable, then whatever limitations to our knowledge may exist will be practical rather than fundamental.

There is evidence to support each of these possibilities. The greatest scientists of the past found it advisable to hedge their bets and espouse both chance and determinism, leaving it to future generations (such as ours) to make a decision as to which was ultimately in control. In this regard it is interesting to note that while Isaac Newton was developing differential calculus, a form of mathematics beautifully matched to the determinism of dynamics, the great Swiss family of mathematicians, the Bernoullis, was constructing the calculus of probability, which is appropriate for games of chance.

I have already suggested that our minds, possibly for genetic reasons related to adaptability and survival, sometimes seem compelled to impose a causal and deterministic structure on nature, leading to the belief that nature is governed by rational law. Regardless of our deterministic mindset, we still recognize that random (i.e., unpredictable) occurrences and chance encounters seem to be embedded in what is an otherwise orderly reality. We purchase lottery tickets in the hope that good fortune will shower us with millions. Even as we play games of chance, however, we pray that undesirable occurrences such as accidents and illness will not befall us.Thus our belief in chance is inconsistent. I, for one, refuse to purchase a lottery ticket, lest the fates forget which particular unlikely event I am hoping for and bestow the wrong one upon me.

If truly random acts do occur, are they amenable to any form of scientific reasoning or mathematical description, or are they by their very nature beyond all hope of understanding? The answer to this question depends in part on how you define a random act. There are some who argue, quite reasonably, that if you dig deep enough, all seemingly random acts are ultimately governed by law. To those believers, randomness is merely the apparently indeterminate, unpredictable, or patternless manifestation of a large collection of causes, each of which is separately deterministic. There are simply so many of these underlying deterministic causes that their net effect is unpredictable or random. In principle, however, the events we call random could be described by deterministic classical physics if we had the time, energy, and computational skill for the task.

Although the belief that determinism can lead to randomness may seem reasonable, it is only partially correct. Some events are legitimately random or unpredictable, regardless of how minutely and with what degree of care they are examined. The microscopic phenomena described by quantum mechanics are intrinsically random. If, as Einstein believed, their randomness is a manifestation of a deeper level of determinism, that level has not yet been found. Indeed, Bell's theorem appears to rule out most reasonable deterministic explanations, leaving some that are even more unappetizing than outright indeterminacy.

I don't think that the existence of randomness at the quantum level would deter the committed determinist, however. Surely the fall of a meteor or the selection of a lottery number does not depend on quantum-mechanical effects, yet we attribute these events to chance also. It is these events, says the determinist, that are really preordained if you dig deeply enough. Let's see if this argument is correct and if quantum phenomena are therefore the only possible case of true randomness, with all other random events being merely superficially so. If this turns out to be so, then the knowledge that is in principle available to us has been greatly expanded. Quantum processes, after all, do not seem to play a significant role in our daily lives. We needn't be overly concerned that we cannot predict the precise time an atom will decay. Accidents, lotteries, illnesses, and other "random" events, however, are quite important to us. It would indeed be satisfying if we had the means of predicting exactly when and how they would occur.

Deterministic Unpredictability

I have some good news and some bad news. The good news is that many important macroscopic processes previously thought to be random are indeed deterministic. The bad news is that these deterministic processes are unpredictable!

How can a deterministic process be unpredictable? Does that mean simply there are so many deterministic processes acting simultaneously that their net effect is, in practice, unpredictable? I'm afraid not. What I am saying is that there are simple deterministic processes that act alone and still produce unpredictable results. Not only don't these phenomena involve quantum mechanics, but they are amazingly simple in classical Newtonian terms. One doesn't have to resort to an argument about a multiplicity of deterministic effects producing an apparent randomness. These phenomena, acting by themselves, produce effects that would meet the randomness criteria of even the most dyed-in-the-wool determinist.

In this chapter we shall study the causes of deterministic unpredictability and its relationship to the quality we call randomness. The entire subject, which has been given the rather forbidding name "chaos," is one of the newest and most exciting areas of physical science and applied mathematics. It has a significant impact on the degree to which we can acquire knowledge about the natural world.

So What Exactly Is Chaos?

Chaos is the name given to a type of process or behavior that is inherently deterministic yet almost completely unpredictable. The current state of a chaotic process uniquely determines its future, but we are unable to say what that future will be. Before I go any farther, let me point out that "chaos" is a mathematical term; the behavior it refers to is a property of certain deterministic mathematical models, which is to say, a property of differential equations and their solutions. Our task will be to go one step farther and connect this abstract property to the observable behavior of certain physical systems. If we can make this connection, we can then say that at least some randomness in nature is rooted in simple deterministic causes. Clearly, not all randomness is deterministic, because, as I already said, microscopic phenomena are inherently random. But certain of the most common macroscopic processes that meet all our criteria for randomness may be describable by simple mathematical models that are deterministic.

Randomness

The meaning of the term "random," which I have so frequently and loosely used, is not generally agreed upon. We have a kind of vague, intuitive sense of what it means. It is the opposite of "lawful," for example. It is also considered to be the opposite of "regular" or "predictable," or of "having a pattern." In short, it seems easier to say what random doesn't mean than what it does. Nevertheless, the concept of randomness is quite important, and mathematicians have recently spent much time and effort in arriving at an acceptable definition of it. I will try to give you some sense of their thoughts on the issue.

To physical scientists, a random process is one that is not governed by any discernible law. The manifestations of a random process have no apparent regularity and therefore occur at unpredictable times, in unpredictable places, and with unpredictable results. Scientists are likely to define a random process simply by giving an example of one. The toss of a coin is considered the quintessence of a random process. The radioactive decay of a single atom is another example. These two processes are quite different; the coin toss is Newtonian, the decay of the atom quantum mechanical. Nevertheless, both are considered to be random processes.

I've heard it said in a news report that a certain killer chose victims at random, meaning that there was no apparent reason for the choice. Of course the killer probably did have some perverse intention that was undiscernible to the public, so that the use of "random" to describe the choice was actually questionable. If the killer simply flipped a coin, however, and shot the first person that appeared after it came up heads, then the killing would be truly random. The great French author André Gide wrote a fascinating story, "Lafcadio's Adventure," about a man who decides to commit a random act of violence by shoving a complete stranger off a train. It turns out that the man he killed was himself a vicious criminal, so that in a perverse way the random misdeed has a good outcome. I mention this only to point out how randomness intrigues us and plays an important role in a world that on a day-by-day basis seems routine and predetermined.

We have already seen that individual microscopic events are inherently random, and there appears to be no law that can describe them with precision. Thus the decay of a single atom is random in principle, whereas the flip of a coin is random in practice. Physicists tend to judge randomness by the nature of the process that produces it, but mathematicians try to discern randomness by examining the outcome of the process rather than its cause. A sequence of ten successive heads (H) produced by flipping a coin would be a random sequence to a physicist because of the way it was produced. If you simply display the sequence, however—H H H H H H H H H H—it certainly doesn't appear to be random in the intuitive sense of the term. If we replace H with the digit 1, then the sequence is 1 1 1 1 1 1 1 1 1 1. If we ask someone if that seems to be a random string of ten digits, I doubt the person would answer in the affirmative. There seems to be a clearly discernible pattern in a string of ten 1s, and patterns and randomness just don't go together.

Mathematicians study strings of digits to see if they can be called random or not. Since the outcome of any repetitive physical process can be represented by numbers—1 for heads and 0 for tails is one example of such a representation—the patternlessness of a sequence should be related to the randomness of the process that forms it. It turns out, however, that judging the randomness of a number sequence is quite difficult. Nevertheless, being able to produce random numbers is important, since they can be used to simulate random physical processes and are also important in certain branches of statistical analysis, such as polling. When pollsters tell us that 46% of the voting population will vote for candidate Jones, they have determined that "fact" by questioning a small, randomly selected sample of the public. If the sample is truly random, their predictions will have a degree of validity, called a confidence level, when applied to the general population. If they took exactly the same-sized sample but did not select it randomly (for example, they polled candidate Jones's family), their results would have no validity beyond the sample itself. It is certainly important that a pollster have an assured method for selecting a group of people in a random fashion.

How then can we tell if a number is random? After considering this problem for many years, mathematicians have recently developed an approach for determining randomness that will be of great importance to us. It is not the only approach, but it gives great insight into both the process by which numbers are generated and the inherent pattern in the numbers.

The method I refer to involves a fascinating concept called "algorithmic complexity." An algorithm is a set of instructions for doing something, usually a computation of some sort. The method you learned in school for doing long division, for example, is an algorithm. The complexity of an algorithm comes from the number of instructions it contains. When an algorithm is converted into a computer program, it is transformed within the computer into a string of binary digits. This word-to-digit transformation may seem rather odd, but obviously the electronic components of a computer cannot read the language of a program directly; it must be converted into a set of instructions that can be "understood" by transistor circuitry. That is what the binary digits can accomplish: the 0s and 1s are represented by low and high voltages. Every algorithm is ultimately represented as a number, which is this very long string of binary digits.

An algorithm's complexity can be reckoned by counting the number of binary digits in the string. Using this concept, the mathematician Gregory Chaitin developed a remarkable theorem: A number is random only if the least complex algorithm that produces it is longer than the number itself. The criterion for the randomness of a number is the length of the instruction set that produces it.

This makes beautiful sense intuitively. If a number has a pattern with a great deal of regularity, you would expect the instructions for forming it not to be very complicated. The regularity simplifies the instructions. If the number seems to have no discernible pattern, such as 1100010110100, you can always instruct the computer to write it with the statement "print 1100010110100." If you express this algorithm as a binary number, it will clearly be longer than the number it constructs, because it contains that very number along with the necessary printing instructions. On the other hand, the number 11111111 111111111111 can always be printed with the algorithm "print 1, twenty times." This algorithm doesn't contain the actual number and can be put into a form that is shorter than it. Mathematicians try to construct algorithms that do more than just print the numbers, but this usually requires a number to be long enough so that whatever pattern it contains can be seen.

There is a fascinating sidelight to this algorithmic definition of randomness. Suppose you have been given a string containing fifteen binary digits and asked whether or not it was random. How many random strings of fifteen binary digits are there? you wonder. In fact, how many strings of fifteen binary digits of any kind are there? The latter question is easy to answer. Since each digit can have only two values and there are fifteen digits, the total number of strings is 2^{15}, or 32,768. According to Chaitin's criterion for randomness, if a digit string can be generated by an algorithm that contains ten fewer digits than the string itself or less, then the string is not random. Actually, if the string can be constructed by an algorithm smaller than the number, it is not random, but requiring a significantly smaller algorithm assures us that the number will have a reasonable pattern. With that in mind, if an algorithm of five digits or less can construct the number, it is not random. How many such algorithms are there? The number of five-digit algorithms is 2^5, or 32; the number of four-digit algorithms is 2^4, or 16; and so forth. Adding them up, we find that there are only 62 possible algorithms that could construct a nonrandom fifteen-digit string. Thus there are enormously more random fifteen-digit strings than nonrandom ones.

Working on probabilities alone, we expect an arbitrary fifteen-digit string to be random. But contrary to expectation, you can't *prove* that the string is random. To do so, you would have to show that none of the 62 algorithms produce the string you were given. "Well," you say, "just run them in a computer and see if the string is generated." Unfortunately, that is easier said than done.

A famous mathematical result says that you cannot prove an arbitrary program will ever stop running and give you an answer. You may get a very small program to stop, but if you had to check a fifty-digit string for randomness, you would be running an enormous number of algorithms, and there wouldn't be enough time left in the universe to check your answer. Although we have a solid criterion for randomness, it does not give us a practical way of generating random numbers.

Deterministic Mathematical Models

Most dynamic systems (those that change in time) of interest to physical scientists, life scientists, economists, and other researchers are modeled mathematically by differential equations. Solving a differential equation requires that you find something when all you know about it is how it is changing. Newton's second law of mechanics, Maxwell's equations of electromagnetism, Einstein's equations of general relativity, and Schrödinger's equation of quantum mechanics are examples of differential equations that model the behavior of particular physical systems. Other differential equations model chemical, biological, ecological, and economic systems. If we are to discuss chaos in a meaningful way, we have to solve some differential equations and consider what they imply.

Until the 1960s, mathematicians thought they understood the behavior of differential equations quite well. Some were so complicated that they could not be solved exactly, but these were relatively few and did not appear to apply to the most common and interesting situations. Besides, even equations that could not be solved exactly could at least be analyzed "qualitatively," meaning that certain general features of their unobtainable solutions could be determined.

The computer enabled mathematicians to attack these previously unsolvable equations using purely numerical methods. High-speed computation provided the first clues that something was seriously wrong with our understanding of how certain differential equations behaved, not just strange and esoteric equations but also fairly simple ones that had been studied (it was thought) thoroughly. Although this is not a book about the methods of solving differential equations, to appreciate the meaning and implications of chaos, it is necessary that we discuss certain of their solutions.

One of the simplest mathematical models that accurately describes a realistic physical process involves the motion of a planar pendulum, a device consisting of a mass m attached to the end of a rod of length L, which swings in one plane around an axle (fig. 11.1).

Depending on how the pendulum is set into motion, it can swing back and

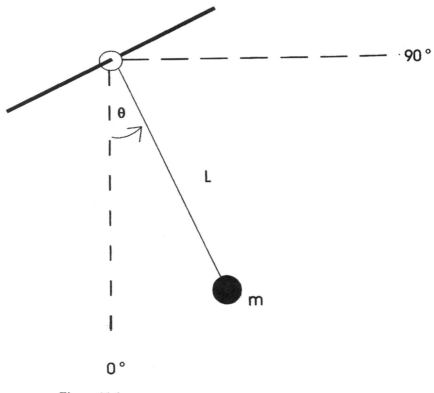

_____ *Figure 11.1*_____
The Planar Pendulum

forth or rotate completely around the axle. Many familiar systems behave very much like a pendulum. The stock market, for example has its "swings" and "rotations." The use of these metaphorical terms is not accidental, since there are many similarities to a pendulum in this trading system. Similarly, the human heart is a muscle that pumps rhythmically like a swinging pendulum.

The simplest treatment of the pendulum restricts its motion to back-and-forth oscillations through small angles, where θ varies between plus and minus about $10°$ from the vertical. When θ is restricted in this way, the force of gravity acts "linearly," meaning that it pulls the pendulum with a force directly proportional to θ. The pendulum becomes important for the study of chaos when it is endowed with three additional features that complicate its motion and make the motion chaotic. The first is friction, which can arise from the axle or from air resistance. Friction is technically known as "damping." The second feature is called "forcing," which is the application of an external rhythmic twisting force (torque) to the pendulum to keep it going. The third feature is a nonlinear force, which is obtained by allowing the pendulum to move through the full range of angles.

If the pendulum is neither forced nor damped and is set into motion by pulling it out slightly and releasing it, it will have its own natural rhythm, called its frequency of oscillation, which is the number of times per second it swings back and forth from its initial angle of displacement under the influence of linear gravity. If the initial angle is small, this frequency can be obtained from a simple formula,

$$T = 2\pi \sqrt{\frac{l}{g}},$$

where l is the length of the rod and g is the acceleration of gravity (9.8 meters per second, squared).

If the pendulum is not damped, it will swing this way forever. If it is damped, the swings will diminish. If the pendulum is given a strong enough initial push, it will rotate completely around the axle. Gravity ceases to be a linear force under these circumstances, so the motion is quite complicated. Nevertheless, in the absence of friction, the rotation goes on forever. If there is friction, the pendulum eventually stops rotating, settles down to a swinging motion, and ultimately stops. The most interesting motion occurs when the pendulum is forced—that is, its axle is twisted rhythmically by an external means (a motor, for example) with some arbitrary frequency that is not usually the pendulum's own natural frequency. The combination of forced motion and friction produces chaotic behavior.

A rather complicated differential equation expresses the changes in θ as time proceeds. When θ is expressed as a function of time, it is written $\theta(t)$ (called "theta of t"). One way to display $\theta(t)$ is a graph in which θ is plotted vertically and t is plotted horizontally to produce a "trajectory." A more informative way of displaying θ is to plot it against the angular velocity, v, of the pendulum, which is the rate at which θ changes. This kind of displacement-velocity graph is called a "phase-space" graph, to distinguish it from the "space-time" graph of displacement time. Figure 11.2 shows a series of θ-versus-v curves, which are called phase-space "orbits." But before discussing this figure, let me tell you a bit about it.

In order to graph all the possible swings of a pendulum, it is necessary to allow the values of θ to range over an infinity of positive and negative values, because the pendulum doesn't just cover limited values of the angle but can rotate continuously in a clockwise or counterclockwise direction. Thus it can go forever back and forth between plus and minus 10° (for example), or it can rotate clockwise from 0° to 360° to 720°, on and on, or counterclockwise from 0° to –360° to –720° (I'm assuming the "0" is at the vertically down position.)

To give you a feeling for the variety of motion of an undamped and unforced pendulum, figure 11.2 shows a set of curves corresponding to swings and rotations of the pendulum. As long as the swings are between about plus and minus 10°, the curves are ellipses (indicated as "1" in the figure). If the swings go through larger angles but do not rotate around the axle, then the nonlinearity of gravity asserts itself, and the curves are no longer ellipses but are still closed (shown as "2" in the figure). At the origin of the axes is a single point, $\theta = 0$, $v = 0$ (shown as "3"), representing the behavior of the pendulum when it is neither displaced nor pushed. Above and below the closed curves are a series of wavy lines, which are the curves of complete rotations (shown as "4" and "5"). These curves are open, meaning that they extend between infinite values of the angle because the pendulum rotates forever. The upper curve is for clockwise rotations, where the velocity is considered positive. The lower curve is for counterclockwise rotations.

Any point in phase space at which both the velocity of the system and the acceleration (the rate of change of velocity) are both zero is called a "critical point." There the system is in equilibrium, since nothing about it will subsequently change. It has no present movement, since its velocity is zero, and nothing to change its movement, since its acceleration is also zero. The point at $\theta = 0$, $v = 0$ is critical. Much less obvious are the infinite number of similar points (labeled "6") at $\theta = + 360°$, $–360°$, $+720°$, $–720°$, and all remaining positive and negative integer multiples of 360°. These all represent the same position of the pendulum (i.e., hanging straight down), but the mathematics demands that the angles have different values.

Another counterintuitive set of critical points is at +180°, –180°, +540°, –540°, and all other positive and negative odd integer multiples of 180°. These are all positions where the pendulum stands on its head, so to speak. The essential difference between the hanging-down and standing-up positions is that the down position is one of "stable" equlibrium whereas the up position is unstable. Stability means that the pendulum will return to its position if it is slightly disturbed, and unstable means that the pendulum will deviate strongly from its position if it is so disturbed. The strangest curves of all are the lobes (labeled "7") that begin and end at the unstable equilibrium points. These curves represent the motion when the pendulum falls from its upside-down position with essentially no initial velocity: it swings around and returns to its starting point.

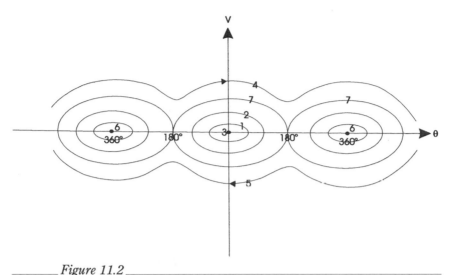

_____ *Figure 11.2* _____
Orbits of an Undamped Pendulum

Poincaré's Contribution

Phase-space orbits sometimes become so complicated that merely examining them visually offers no insight into the nature of the motion that produced them. The French mathematician and astronomer Henri Poincaré had a brilliant idea for eliminating some of this complexity. If the phase-space motion is observed only at certain specific times, the orbit becomes a series of points, each point being the displacement-velocity configuration of the system at the specific time of observation. It's like shutting off the lights and turning them on only for brief instants of observation, or taking a series of snapshots rather than a complete motion picture. In the case of a pendulum, if the times of observation are separated by a period equal to the time the motion takes to complete a full cycle, then the phase-space point always appears at the same place in the θ and v dimensions. If all points of intersection are drawn on a single plane, it gives a picture of where a system is at many times, called a "Poincaré section." The pendulum's section contains just one point, so it is certainly a simpler representation than the full orbits of figure 11.2.

The pendulum that we have been examining is called a conservative system because its energy is conserved (does not change). In the real world, the mass and rod are subject to a small but noticeable amount of frictional resistance due to both the ambient air and the axle. Because of this damping, the pendulum's mechanical energy is converted to heat, there is no longer energy conservation (the heat energy must be considered lost), and the system changes from a conservative to what is called a "dissipative" one. Dissipative

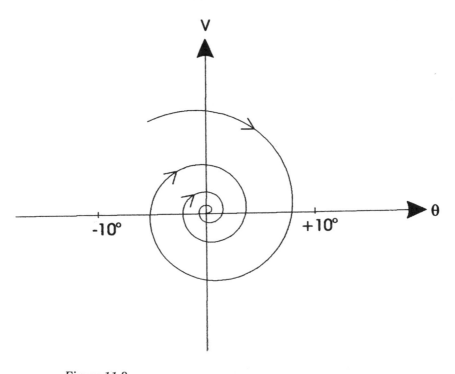

Figure 11.3

Spiral Orbit of a Damped Pendulum

systems are far more common in nature than conservative ones, because friction of one type or another, as well as other mechanisms for losing energy, is always present.

We will examine the θ-versus-v relationship for the back-and-forth motion through limiting angles (the orbits that were ellipses in the absence of friction, numbered "1" in fig. 11.2) of a damped pendulum in one example of a new orbit. The phase-space solution is now a spiral, as shown in figure 11.3. There is still an equilibrium point (critical point) at the origin, which is not at all surprising since even a damped pendulum will do nothing if it is not set into motion initially.

The spiral orbit appears to terminate at the origin, but this is not so. It takes an infinite amount of time for any orbit to reach the origin, corresponding to the infinite time it would take the pendulum to stop moving completely, once it is set into motion. In this mathematical idealization, the origin is approached but is never actually reached. The spiral orbit and the critical point are separate and distinct.

This is one example of an interesting property of the differential equations that describe nearly all physical systems. Different orbits of such equations can never cross each other, nor can they even touch each other. If two orbits did cross each other, the point of their crossing would represent a situation in which the system had the same position and velocity yet was moving along either of two different physical paths. But it can be shown that the solutions of these equations are unique, meaning that an orbit is uniquely determined once a position and velocity on it are known at a given time. Therefore it would be impossible for two different orbits to have such a point in common.

This is precisely what is meant by saying that a system is deterministic. Its future motion is uniquely determined once the values of its position and velocity are known at any instant of time. A system that could equally well move along two or more different orbits, starting from the same position and velocity, would be a nondeterministic system. It is certainly possible to create differential equations that model nondeterministic systems, just as it is possible to create drugs for diseases that do not yet exist, but this is not what we are interested in doing at present.

The critical point at the origin, toward which all spiral orbits tend but which none ever reach, is called a "limit point" and also, even more descriptively, an "attractor." Recall that the undamped pendulum also had a critical point at the origin, but the elliptical orbits surrounding it kept their distance. Therefore it is not an attractor. Knowing that a certain system has attractors and—better yet—knowing their locations in the phase space of the system are invaluable bits of information that do not require the differential equations to be solved. Mathematicians quite naturally wondered if critical points were the only types of attractors that could occur. The answer, as I will now demonstrate, is an emphatic "No."

Nonlinear Oscillators, Attractors, and Limit Cycles

As physicists began to study oscillating systems that were even more complicated than the damped-pendulum system of the previous example, they found themselves confronted by forces that were not as simple as the linear-gravitational force that makes a pendulum swing through small angles or the force of air resistance that slows the pendulum down. The planar pendulum we have been looking at can provide us with the kinds of more complicated forces I am referring to. If even this "simple" pendulum is set into motion by making it move through a broader range of angles or by giving it so much energy that it rotates completely around its axis, then the force of gravity does not behave as an approximately linear force that depends on the angle, but as a full-fledged nonlinear force that depends on the sinc of the angle. The effect

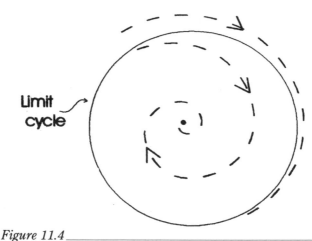

_____ *Figure 11.4* _____

Pendulum Clock Limit Cycle and Orbits

of this nonlinearity is to change the closed orbits from ellipses to more complex shapes and also to produce a series of waves that represent rotational motion. We have already seen this in figure 11.2.

Limit Cycles

When a planar pendulum is damped, some of its infinitely numerous equilibrium points become attractors. There is a class of systems whose attractors are not points but closed orbits. Such attracting orbits are called "limit cycles." A common example is a pendulum clock. After the clock's spring mechanism is wound, the pendulum must be pulled out slightly and released. If you do not pull the pendulum out sufficiently, the spring mechanism is not activated, and the pendulum swings back and forth a few times and eventually stops in the vertically downward position. That position is an attractor for all initial displacements that are too small to activate the mechanism. If, on the other hand, you pull the pendulum out farther, it swings back and forth a few times and settles into a periodic oscillation that is perpetuated by the spring mechanism. The periodic motion is a limit cycle that is an attractor for all large initial pendulum displacements. Figure 11.4 shows a phase-space diagram consistent with the pendulum's different types of motions.

In order for a system to have a point attractor, it must be able to lose (dissipate) energy. Systems that conserve energy can engage in periodic motion, but they cannot be attracted from states where they have energy and are in motion to a single state where they have no energy and there is no motion. On the other hand, a system that is damped and consequently dissipates energy cannot by itself maintain a periodic orbit, since the loss of energy causes the orbit to decay. The only way a system can have an orbit that is both periodic and an attractor is if it has one mechanism for dissipating energy and another for receiving energy. A stable periodic orbit can result, in which the energy that is fed into the system is exactly dissipated. Such a system is said to be both forced (given energy) and damped (made to lose energy). In a pendulum clock, the forcing is accomplished by the spring mechanism, which doles out to the swinging pendulum at the correct instants of time the energy it was given when it was wound. The damping is due to air resistance, other forms of friction, the generation of sound, and mechanical vibrations set up in the clock casing.

If the system is forced and damped and has nonlinearities, and if it is sufficiently complex in its structure to have at least three different types of behavior (degrees of freedom) that are capable of being produced by the delivered energy, chaos can result. Although this may seem to be an extremely restrictive set of conditions, it turns out that most real systems satisfy it. A pendulum, for example, can swing back and forth and rotate in two different directions.

I can give no simple mathematical explanation for these conditions for chaos. Such a system will try to behave in a way that allows the energy being fed into it by the forcing mechanism to move through various degrees of freedom and be dissipated by the damping mechanism. Thus the system can be viewed as a conduit for dissipating energy. Because the system is also nonlinear, as energy flows through it, its motion can acquire sudden exaggerated irregularities as first one and then another of its degrees of freedom are activated. This appearance of behaving randomly is a hallmark of a chaotic system.

The Forced, Damped Planar Pendulum

The forced, damped pendulum makes a particularly good example for illustrating how chaotic motion arises. As simple as this system may be mechanically, however, its mathematics is so complicated that the full range of its motions must be described by means of computer-generated numerical solutions (R.L. Kautz, *Amer. J. Physics* 61[5] [May 1993]: 407).

I will briefly discuss the behavior of this system under a few different conditions and describe some of its corresponding orbits. First, we must think about how a pendulum can be forced. If you look at fig. 11.1, you can imagine a mechanism for applying a rhythmic twist—a torque—to the axle on which the pendulum swings. If we were to build such a pendulum, we would need some kind of mechanical contrivance that would allow us to apply a torque to the axle with an adjustable magnitude and period. We couldn't simply affix a motor to the axle; we would need to attach the motor with some sort of frictional grip or clutch mechanism that would allow it to exert no more than a required amount of force on the axle and would allow the axle to slip if the pendulum exerted a greater countertorque. It is much simpler to model this torque mechanism on the computer than to actually build it.

Such models tell us that the behavior of the pendulum depends on both the period and the strength of the external torque, which we control, relative to the forces of damping and gravity, which are fixed by the structure of the system. I will divide my description into three parts, beginning with the external torque at zero (no torque at all), increasing to a torque that is weak relative to gravity, and finally increasing yet again to a torque that is strong relative to gravity. The behavior of the pendulum changes radically. In the final case of strong torque, I will also discuss another type of change, varying the period of the torque, its rate of twist. As the period is varied, the onset of chaos is seen. In other words, forcing a damped pendulum to oscillate at certain rates and with a certain magnitude of external torque produces apparently random and unpredictable behavior. Yet the behavior is being generated mathematically by a simple, deterministic computer program.

No Driving Torque

Suppose, first, that there is no external torque. Two kinds of motion are possible, depending on the initial energy given the pendulum. If the pendulum is displaced slightly from the vertical (giving it gravitational energy) and released with little or no velocity, it will oscillate back and forth, ultimately dissipating its energy and coming to rest. We have already seen the phase-space diagram of this damped oscillation, a spiral that tends to the point attractor at the origin (see fig. 11.3).

If the pendulum is given sufficient initial energy to make it rotate about its axis in one direction or the other, it will swing around a number of times until its energy is insufficient for continued rotations and then settle into oscillations that gradually bring it to rest.

Weak Driving Torque

If the driving torque is weak compared to the gravitational torque, the fixed point attractor in figure 11.3 disappears and is replaced by a periodic attractor or limit cycle that is similar to that of the pendulum clock shown in figure 11.4.

The period of this motion turns out to be the same as the period of the torque. The pendulum settles into the limit-cycle orbit shown in figure 11.4 only after it has gone through some rather arbitrary initial motion. It will not execute the limit cycle from the very beginning, unless, by pure luck or careful calculation, you give it an initial velocity and displacement that are right on the cycle. This situation, as I have said, is very much like a pendulum clock, except that the clock's escapement mechanism applies torque in the form of discrete impulses synchronized with the natural frequency of the pendulum.

Strong Driving Torque and Basins of Attraction

If the magnitude of the driving torque exceeds that of the gravitational force, the single periodic attractor is replaced by three attractors. The first of these is simply another periodic attractor, like the clock. The other two, however, are quite different, and their appearance is surprising. One is a complete, continuing clockwise rotation, the other a similar counterclockwise rotation. It is certainly unusual to see a periodic torque, whose direction of rotation keeps changing, produce an attractor that rotates in one direction only. This demonstrates the fact that even conceptually simple systems can produce counterintuitive motions, given the right conditions.

The particular attractor that the pendulum is ultimately drawn to depends on its initial displacement and velocity. It is instructive to divide phase space into three regions, each of which contains the initial conditions that lead to one particular attractor. These regions are called "basins of attraction." One of the features of chaotic systems is that as system parameters (such as magnitude and frequency of the external torque) are varied, the number and type of attractors can increase dramatically. These fundamental changes in the nature of the resulting motions are called "bifurcations," a term introduced by the mathematician C. Jacobi in 1884.

The appearance of each additional attractor brings with it an additional basin of attraction. These regions of phase space tend to become extremely convoluted in shape and to interweave with each other to the point where it becomes almost impossible to separate one region from the other on a diagram. The basins are so tightly interwoven that two sets of seemingly identical initial conditions, the same point drawn twice, might belong to different basins, with the pendulum being drawn to significantly different attractors. The basin diagram constitutes what is known as a "fractal," a strange geometrical object that plays an important role in chaotic systems. We shall discuss some of its properties a bit farther on.

I intimated earlier that nonlinearities in a system seem to account for the apparently random and exaggerated motions of chaotic behavior. The interweaving of basins of attraction accounts for yet another of the characteristics of chaotic systems, the fact that they behave differently from each other even though they are set into motion with what appear to be the same initial conditions.

This makes such a system seem unpredictable, because we cannot say what it will do the next time we start it, even though we observed its actions previously when we started it in what we thought was an identical fashion. Today's behavior provides no indication of tomorrow's.

The system would behave in exactly the same way if we could truly start it out with exactly the same set of conditions. The dense interweaving of basins of attraction prevents us from doing this. No matter how precisely we think we have chosen our initial conditions, no matter how carefully we have set the system into motion, we will inevitably be off by an "insignificant" amount that nevertheless could place us in any one of several basins of attraction. This unpredictability of the future behavior of a system, which is also called "extreme sensitivity to initial conditions," is another hallmark of chaos. We shall now discuss yet another chaotic property, the incredible complexity of the system's attractor, producing a form of motion that appears to have no comprehensible structure. This is related to the exaggerated motions created by nonlinearities and the passage of energy through different degrees of freedom.

Period Doubling

If the driving frequency is decreased below the natural frequency of the pendulum, another interesting phenomenon, called "period doubling," occurs. In this situation, the pendulum executes two different orbits during two successive periods of the forcing torque. We might call them orbits A and B. Such a double orbit can be viewed as a single orbit whose period is twice that of the torque or as two different, alternating orbits, each of which has the period of the forcing torque. If you chose to watch the pendulum only during every other period of the torque, you would see just orbit A or orbit B. If you constructed a Poincaré section that intersected the double orbit at a set of times that differed by the period of the torque, there would be two points on the section, one contributed by A and the other by B. If you chose to have the sections intersect the orbit at times that differed by two periods, there would be only one point on the section, corresponding to A or B.

If the period of the external torque is gradually reduced, additional period doublings occur. Thus we next observe a period-four solution, followed by a period-eight solution, and so on. Each doubling is a bifurcation. If the successive doublings are visualized by constructing Poincaré sections at intervals of a single period, the sections will show two points, four points, eight points, and so forth as each new doubling occurs. If we watched the system with a strobe light that illuminated it with the period of the torque, we would see a succession of different positions and not realize that the system is exercising an orbit at a period other than the one we are observing it with.

Strange Attractors

Until quite recently, it was believed that there were only three types of attractors associated with the kinds of differential equations that model physical systems. We have been considering two of them, the limit point and the limit cycle. The third type, which I will merely mention because it will be of no particular importance to us, is called a "quasi-periodic solution" of the differential equation, and it can be pictured as a curve that winds around a torus (a doughnut-shaped surface) in phase space.

These three types of attractors were easy for mathematicians to visualize and describe. Then a fourth type was discovered whose existence has led to a revolution in the way we think about the mathematical models of even the simplest physical systems. It is called a "strange attractor," and its strangeness lies in the fact that it is a curve of such unbelievable properties that it cannot even properly be called a curve. It is so twisted and convoluted that no pen or pencil could ever hope to draw it. In mathematical terms, it is a curve of "fractional dimension," now called a "fractal." A limit point, being a true point, is said to have a dimension of zero. A line, even when bent into a curved limit cycle, is one dimensional. A plane is two dimensional. But a strange attractor can have a dimension between one and two (or between two and three, etc.). It is more than a line but less than a plane, or more than a plane but less than a three-dimensional solid.

Historically, the first appearance of a strange attractor can be attributed to the work of the meteorologist E. N. Lorenz, who published his observations in 1963. Lorenz was studying a set of three coupled differential equations as a model for the motion of the atmosphere under severely simplifying assumptions.

We now know that a surprisingly large number of systems have strange attractors under the right conditions. We will even find such an attractor in the forced pendulum we are now studying as we vary the period of the torque to produce a succession of doublings. The end result of these doublings when we approach an infinite number is the strange attractor. We can think of it as representing an "orbit" that has doubled and redoubled so often that it defies pictorial representation. However you choose to think of it, there is simply no meaningful way to draw this attractor in phase space. The Poincaré section now becomes a necessity rather than a convenience.

Every point on the section is the position of the orbit at a single instant of time, the instants being separated by exactly one period of the forcing torque. If the cycle had a single period, the section would have a single point. But each successive doubling of the period puts another point on the section at a different position. The diagram that emerges after an infinite number of period doublings is a fractal.

Fractals possess a feature called "self-similarity," which means that their structure looks the same at whatever scale it is displayed. If you could examine a very small segment of it under a microscope, you would see the same diagram that you see with the naked eye. This chaotic structure manifests itself to us as randomness, even though it actually possesses a complex and individualistic order. If you could observe the pendulum moving continuously through a significant portion of its orbit rather than seeing just the Poincaré section, the motion would seem totally disorganized, with no discernible pattern or order.

The basins of attraction of two rotating pendulum orbits also form a fractal. That fractal does not evidence the same type of chaotic motion that results from the fractal nature of the strange attractor. The strange attractor is a single limiting orbit with unusual qualities, whereas the basins of attraction depict highly interwoven regions of phase space that lead to different but unexceptional limiting orbits. Yet in practice, the fractal basin structure is also associated with effects that are virtually indistinguishable from true chaos. Because the regions are so inextricably intertwined, initial conditions that seem identical to us can easily reside in different basins. Two pendulums can be set in motion in "exactly" the same way, yet perform dramatically different ultimate motions. One can rotate clockwise, the other counterclockwise. Although the final motion is orderly, the two different behaviors seem to occur for no reason.

In the following section, I will discuss the emergence of chaos in a discrete model, also called a "discrete map," where the mathematics is simpler, the description of a process that does not occur continuously in either space or time. Using such a model, I can more effectively explain what I mean when I say that chaos is indistinguishable from randomness. The only problem with using a discrete rather than a continuous model is that the system being modeled will not be one that we are familiar with. The pendulum's virtue is that we have all seen one. Its vice is that the mathematical model that describes its behavior is a differential equation, and few of us have ever seen one of those, much less solved one. When the system is familiar, its model is difficult; when the model is simple, the system is difficult.

Discrete Maps

Poincaré introduced an element of discreteness into the analysis of continuous motion by suggesting that it could be studied more easily by observing it at certain specific instants of time. The development of the computer some fifty years later introduced a second element of discreteness into the study of continuous phenomena. Digital computers do not do continuous mathematics; they do discrete mathematics. The differential equation, which forms the basis of all the models we have been discussing, is a construction of the human mind, which seems to operate in a continuous rather than discrete mode. To get a computer to solve a differential equation, the equation must first be converted into what is called a difference equation. This means that the infinitesimal differences represented by differentials, which must be thought of as being of "zero" size, are replaced by differences that, although small, are finite. Of course even finite differences can be made very small, but the limit on computer memory prevents them from ever being truly infinitesimal like a differential.

A difference equation is simply a relationship between the value of a quantity at one time and its value at another time that differs by a small but finite amount. In the jargon of mathematicians, a difference equation is a "finite mapping," meaning it generates finite changes in a variable. The Poincaré map doesn't really "generate" finite changes; it merely displays them, because it depicts a legitimately continuous change at a succession of discrete times. Yet from the standpoint of a mathematician, a succession of points can be thought of as a map, regardless of the underlying scheme that produces it. We see then that the study of finite mappings should be of interest, both because such mappings represent the solutions of difference equations and because they furnish a novel way to study complicated continuous motions.

The Logistic Equation and the Logistic Map

Rather than deal in generalities, I will choose a specific differential equation of great importance and then examine its associated difference equation. My choice is the logistic or Verhulst equation, which is taken from the field of population dynamics. This equation was originally proposed in 1845 by Pierre-François Verhulst, a pioneer in that field, as a model for the growth of a population in a limited environment. The limits could refer to resources or the ability of the environment to eliminate waste and pollution. Verhulst treated the population as though it were composed of individuals whose birth rate is linear (so that increase depends only on the first power of the population) but whose death rate is quadratic (so that decrease depends on the square of the population).

These mathematical assumptions have interesting physical corollaries. A birthrate that depends on the first power of the population implies that births are a result of individual activity. A death rate that depends on the second power of the population implies that death is the result of an interaction between an individual and the individual's neighbor. The power (i.e., the exponent) of the interaction reflects the number of individuals that come into contact. Such a model is not unreasonable for populations that reproduce asexually (e.g., certain bacteria) and that tend to die off because they are sensitive to overcrowding.

In his model, Verhulst summed up all of these factors by making the death rate quadratic. In the Verhulst equation,

$$\frac{dx}{dt} = \beta x - \delta x^2,$$

the variable x represents the population at a time t, βx is the linear birthrate (births per year), giving the population's rate of increase, and δx^2 is the death rate, giving the population's rate of decrease. It is a differential equation, since it is seeking dx/dt, the rate of change of the population (the net growth rate), which is a derivative, the ratio of a small population change, dx, to a short period of time, dt, in which that change occurs. The equation is often written in a slightly different form,

$$\frac{dx}{dt} = \beta x(1 - \frac{\delta}{\beta} x),$$

in which it is easy to see that the population does not change in time when either $x = 0$ or $x = \beta/\delta$. In the jargon we have already used, the values of x that make the right-hand side of the equation zero are critical points of the equation and equilibria of the population. The $x = 0$ value makes intuitive sense; if there is no population, it cannot change because there will be neither births nor deaths. True, but not very interesting. The other equilibrium value, however, is quite interesting, because it represents a stable nonzero population, for which the birth and death rates are exactly balanced. Demographers call this situation zero population growth, and ecologists call the value β/δ the carrying capacity of the environment in which the population lives.

To solve Verhulst's differential equation numerically, it must be converted to a finite difference equation. The infinitesimally small change in population, dx, is converted to a finite difference, $\Delta x = x_{t+\Delta t} - x_t$, which is the difference between populations at two neighboring times, $t + \Delta t$ and t, separated by the finite duration Δt. The values of x at these particular values of t are designated by subscripts, $x_1, x_2, \ldots x_n$. The differential equation becomes the difference equation:

$$x_{n+1} = r x_n (1 - x_n),$$

where r is a constant that arises from the details. Since the difference equation is just an approximation to the logistic differential equation, which has a well-behaved solution, we would expect that it also to produce a well-behaved sequence of numerical values, $x_1, x_2, \ldots x_n$. We shall see that this is far from the case: the difference equation, or discrete map, displays a fascinating and unexpectedly complex behavior that the differential equation does not.

Continuity or Discreteness?

Before tackling the difference equation, I will make one comment about the importance of this whole exercise. If the difference equation is viewed as an approximation to a correct differential equation, then any unusual behavior it might exhibit could be blamed on the faults inherent in approximations. However, it is equally reasonable to view the differential equation as an approximation to the difference equation. Differential equations assume the existence of continuous behavior and are capable of modeling only such behavior. Is the physical world really continuous though? If we desire to model the dynamics of population growth, we must ask ourselves if a population is a continuous sort of thing. When the population is small, such as a family of two, and it adds a child to produce a population of three, the jump from two to three is discontinuous. It would be meaningless to describe that kind of growth with a differential equation, but a difference equation might be eminently reasonable. On the other hand, if it is the population of the entire world that is being described, then the increase on an hourly or daily basis could quite reasonably be described as being smooth and continuous and thus suited to being modeled by a differential equation.

In short, if births and deaths can be considered as occurring constantly, their impact on a population is continuous. This is the case in a very large population. If it is more reasonable to view the births and deaths as occurring at discrete times, then the process is not continuous, and it is the differential equation that is only an approximate model. Certain insect populations, for example, live for only a year. During this time they lay the eggs that will be responsible for the population in the following year; then they die. Describing the population dynamics of such insects requires a discrete map rather than a differential equation. Thus it is important to study discrete maps in their own right and not as approximations to some other model.

The Behavior of the Discrete Logistic Map

While the logistic map could be analyzed numerically on a computer, there is an alternative analysis that demonstrates its unusual behavior pictorially and with much greater clarity. Think of the map as a function that takes as its input the value x_n and produces as its output the value x_{n+1}. We can represent this property on a graph that relates the two successive values of x. In such a graph, the axes are x_n, the abscissa, and x_{n+1}, the ordinate. The graph itself is the function $x_{n+1} = rx_n(1 - x_n)$, which is a parabola whose abscissa

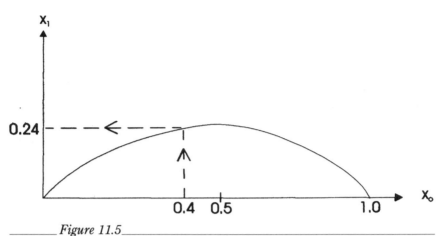

_____ *Figure 11.5* _____
Parabolic Mapping

intersections are at 0 and 1, whose peak is at 0.5, and whose peak height (ordinate value) is $r/4$.

To generate a sequence of x values, one would first draw the graph with x_0 as the abscissa and x_1 as the ordinate (fig. 11.5). A particular value of x_0 would then be chosen (the initial value for the problem) and the resulting value of x_1 read directly from the graph as shown below. Suppose we set $r = 1$ and choose $x_0 = 0.4$. As indicated in figure 11.5, this produces the value $x_1 = 0.24$.

Following this first iteration of the map, the axes would be relabeled, with x_1 now playing the role of abscissa and x_2 becoming the ordinate. The previous ordinate value, $x_1 = 0.24$, now becomes the abscissa value, and a new ordinate value, $x_2 = 0.184$, would be read off the graph. We continue making the previous ordinate the new abscissa, always retaining the same graph, and in that way we can generate as long a sequence as we want.

This approach, although correct, is tedious. It is admirably suited to the computer, which never seems to get bored with its task. There is an alternative approach for impatient humans, however, which allows you to see the generation of all the x values on a single graph, shown in figure 11.6. This approach requires only an additional line on the graph, $x_{n+1} = x_n$ (a straight line with slope 1), which is simply a function that says "ordinate equals abscissa." The entire sequence of iterations can be drawn on a single graph by bouncing a dashed line between the parabola and the line $x_{n+1} = x_n$, which has the same effect as constantly relabeling the axes and transferring the old ordinate value to the new abscissa.

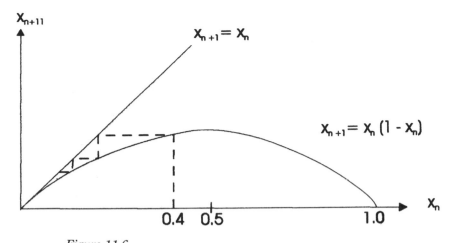

_____ *Figure 11.6*_____
Parabolic Mapping and Iterations

Looking at the graph, we can now see that if we continue bouncing the values between the parabola and the line, we ultimately approach the value 0. This value is ultimately reached regardless of the initial value that we choose. In real terms, this means that the population being described by this particular value of the parameter r is doomed to extinction. The $x = 0$ point is the analog of a critical point of a differential equation. In the present case of a discrete map, the terminology changes somewhat, and it is called a fixed point. Not only is 0 a point of equilibrium, meaning that the population stays there once it gets there, but it is also an attractor. Every initial population is inevitably drawn to 0.

Armed with this technique for visualizing an entire sequence of x values on a single graph, we can begin to examine such sequences for different values of r. Suppose, for example, we increase r from 1 to 2. The parabola rises, and the reflection line now intersects it at $x = 0.5$. If we now choose an arbitrary initial value of x, say $x_0 = 0.40$, we see that the sequence of subsequent x values now heads toward $x = 0.50$. Whatever initial value we choose, with the sole exceptions of 0 and 1, the sequence of successive populations is attracted to the point $x = 0.50$. That point has obviously become the new fixed point, and it too is an attractor. The earlier fixed point, $x = 0$, is still a fixed point, but in a much more restricted sense. If $x = 0$ is chosen as an initial value, it will remain at $x = 0$. But any initial choice that is even infinitesimally distant from $x = 0$, such as $x = 0.000001$, will move toward the fixed point $x = 0.5$. We call $x = 0$ an unstable fixed point and also a repeller.

A Bifurcation Occurs

Suppose we increase r once more, this time to a value greater than $r = 3$. Instead of the sequence x_n approaching a single fixed point, as was previously the case, an infinitely repetitive cycle forms, in which two different limiting values are approached for alternating values of n. We call this a stable cycle of period 2 (the period referring to the number of iterations between successive values of the same limiting point). Physically, this would correspond to a population that varies in size from one "year" to the next. (Of course the time is not a year; it is whatever value is assigned to Δt.) Such behavior is seen in the populations of certain insects, like gypsy moths.

This sudden and drastic qualitative change from a single fixed point to a stable cycle as the parameter r is varied is another bifurcation, just like the changes we met when studying pendulum motion. Bifurcations, which occur in differential equations as well as discrete mappings, always involve a fundamental change in the nature of a solution as a parameter is varied. Quite often the previous fixed point does not disappear but becomes unstable and no longer plays an important role in the behavior of the physical system.

Constructing a Bifurcation Diagram

The graphical approach that we have been using is quite valuable for seeing how and why different types of limiting behavior appear. For a detailed study of the mapping, however, we have to examine an enormous number of r values, and this is a job for the computer. Therefore we will temporarily put away our parabola and discuss how a computer analysis would proceed and what it would show us.

A properly constructed computer program would generate a graph of limiting x values versus values of r. It could, for example, allow us to input values of r between 0 and 4 in steps of 0.01. For each r value, the same initial x value is selected, and the program proceeds to iterate the mapping a selected number of times—fifteen hundred, let's say. The final thousand of these iterations are then plotted on a vertical axis directly above the r value associated with them. The first five hundred values of x are generally discarded, as they are considered to represent the transient behavior of the mapping rather than the limiting behavior, which is what we want. For those r values that correspond to a fixed point (r between 0 and 3), the final thousand iterations will be identical and will appear as a single point on the graph. Period 2 cycles will appear as two points. As the values of r are successively dealt with, the computer is actually drawing a bifurcation diagram for the mapping. Figure 11.7 shows a computer-generated bifurcation diagram for a range of r values between 0 and 3.57 that produce a single point as well as period 2 and period 4 cycles.

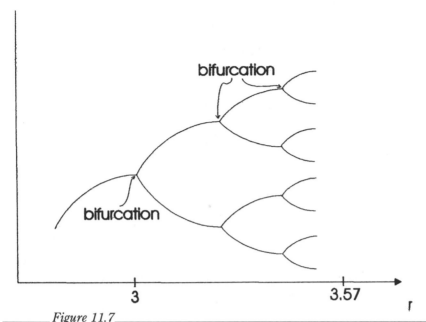

_____*Figure 11.7*_____
Bifurcation Diagram

If the value of *r* is allowed to increase beyond *r* = 3, a succession of period-doubling bifurcations occurs in which the period 2 cycle becomes a period 4 cycle, which in turn becomes a period 8 cycle and so on, ad infinitum. This so-called period-doubling cascade continues until the value *r* = 3.57 is reached, with the range of *r* values for each successive cycle becoming smaller.

The strangest behavior of the map is yet to come. As *r* is allowed to move past the value 3.57, an enormous number of sequences (discrete orbits) appear that do not seem to have a fixed point or a stable cycle of any period. The last thousand *x* values at these values of *r* scatter all over the graph in the vertical direction. These apparently nonperiodic orbits are called "chaotic orbits."

The thousand points that now fill the vertical direction of the bifurcation diagram can be thought of as representing a kind of attractor, because whatever initial value of *x* we have chosen, the final thousand iterations produce a virtually identical set of points, although in a different order. We call this a strange attractor, being somewhat at a loss for words. It can be shown that some other initial *x* value, as close as you like to the value that produced this particular orbit, would scatter its last thousand iterations among this same set of points.

Unfortunately, there would be virtually no relationship between the particular placement of (let's say) the 1,201st iteration of one initial value and the same iteration of the other. The only thing they have in common is the fact that they occupy the same attractor. On average, the distance between corresponding iterations will be on the order of the width of the entire attractor (approximately 1), whereas the separation of the initial values could have been as small as the computer program would permit. This, as we shall see, is the essence of chaos. Keep initial values as close as you can, yet watch the ultimate values deviate by as much as they can! Chaos can be thought of as a property of behavior on the strange attractor that destroys any memory the system may have had of the closeness of its initial values.

The region between $r = 3.57$ and $r = 4$ (not shown in the figure) is not made up entirely of chaotic orbits or strange attractors. As the computer program moves slowly (by steps of 0.01) through its range of r values, it discovers narrow regions (called windows) in which stable cycles once again appear, islands of order within the sea of chaos. Unlike the periods of 2, 4, 8, and so forth that characterized r between 2 and 3.57, these windows may display cycles with any period. Moreover, if one of the windows is scrutinized more closely, for example by having the program increment r in steps of 0.0001 within it, the structure within the window looks exactly like a miniature version of the entire bifurcation diagram that contains the window. This is another example of self-similarity and fractals.

An entire book could be written about the complexities of structures in the bifurcation diagram of a logistic map. The appearance of chaotic orbits is merely one symptom of the incredible mathematical pathologies lying hidden within a map that pictorially is nothing more than an inverted parabola.

The Nature of Chaos at r = 4

The values of r between 3.57 and 4 produce both chaos and simple periodicity. At $r = 4$, the chaos is most pronounced because the attractor fills the region between 0 and 1 in the vertical direction, which it does not do at any other r value. An even more significant feature of $r = 4$ is that the mapping can be solved exactly, without recourse to the numerical computer solution. This exact solution allows us to see more clearly what the implications of chaos are for the origin of randomness in nature.

Because the solution is difficult to calculate, I will simply present it prefaced by the notorious phrase "It can be shown that." In fact, it really can be shown (see Joseph Ford, "Chaos," in *Chaotic Dynamics and Fractals*, ed. Michael F. Barnsley and Stephen G. Demko [San Diego: Academic Press, 1986], p. 4) that the logistic equation with $r = 4$ is equivalent to the much simpler equation

$$x_{n+1} = 2x_n \pmod 1,$$

which says: in order to obtain x_{n+1}, you must double x_n and throw away the integer part in front of the decimal point (which is what the "mod 1," short for "modulo 1," means). This new mapping can be stepped backward to produce an exact solution for x_n in terms of x_0:

$$x_n = 2^n x_0 (\text{mod } 1).$$

The implications of this simple result are enormous: any iteration of the logistic map is obtained from whatever initial value was chosen by multipying the value by an integer power of 2. You may wonder why such a simple and deterministic result should lead to chaos. To answer this question, let's express the initial value, x_0, as a binary rather than a decimal fraction. After all, the values of x must lie between 0 and 1 (because we throw away the integer part), so x is indeed a fraction.

A binary fraction, which is much less familiar than a decimal fraction, consists of a 0 and a point (we can't call it a decimal point) followed by a string of 0s and 1s, such as 0.0101. Each position in the string, whether to the left or right of the point, represents a power of two, and the 0 or 1 in that position is the multiplier of the power. Positive powers (and the 0 power) lie to the left of the point, negative powers to the right. The value of the fraction represented by this binary string is the sum of all the powers: $0.0101 = 0(2^0) + 0(2^{-1}) + 1(2^{-2}) + 0(2^{-3}) + 1(2^{-4})$, which as a decimal fraction is 5/16, or 0.3125. Multiplying a binary number by a power of two is a simple operation: move the point to the right as many places as the power. Thus $(2^3)(0.0101) = 0010.1$.

We see that the logistic map at $r = 4$ simply shifts the point of the initial value as many places to the right as the map is iterated and then discards whatever is in front of the point. If you want a hundred iterations, shift the point a hundred places to the right and get rid of the 0s and 1s to the left of the point. This is a simple deterministic algorithm. The future (the hundredth iteration) is completely and uniquely determined by the present (the initial value).

Then, I repeat, where does the chaos come from? Well, I answer, how many places does your initial value have? Think about this for a moment. The number you plug into your program as the initial value has to come from somewhere. There are two possibilities. Either it represents a real measurement you made on some population—let's say you counted all the gypsy moths in a one-acre lot—or else you plucked it out of a hat and typed it into the computer. If you counted moths, then the number of places is determined by the accuracy of your count. If you typed in an arbitrary number, it is determined by the precision of your computer.

In either case, you are about to run into a serious problem. Suppose you invented the number, and your computer will allow you sixteen places to the right of the point. By the sixteenth iteration, your result is using the last bit of data that is genuine, that is, part of your initial value. If you want seventeen iterations, you are out of luck as well as meaningful information. Everything you generate beyond the sixteenth iteration, no matter what it looks like, is simply noise produced by the computation process (the software and your program) or by the computer itself (the hardware). Suppose you selected two different initial values that differed only in the tenth place and beyond. By the eleventh iteration, any semblance of similarity would be lost. From the eleventh iteration onward, the only part of the initial information that matters is what these values were *beyond* the tenth place, which was of no concern to you when you chose them. The chaotic process eventually magnifies differences that are so minute as to be beyond your control and perceptive powers.

The same thing happens to initial data that you measure. Suppose your count of gypsy moths is accurate to six places, and you are concerned about their populations in ten years or more. Forget it! By the sixth year, if your count was off by even one moth, the error is more important than the thousands of moths you counted correctly. This is the origin of the so-called butterfly effect, which states that in order to increase the accuracy of a weather prediction by more than a few additional days, you would have to take into account effects on the motion of the atmosphere as small as the beating of a butterfly's wings. The hope of improving the accuracy of forecasts with bigger and better computers, more stations for collecting data, satellite observations, and the like is probably doomed to failure because the development of weather patterns is described by equations that appear to be fundamentally chaotic.

We can now begin to see why determinism and predictability are independent. Determinism means that a final value can be uniquely determined from an initial value. The logistic map gives you that. Predictability means that you can use the determinism to make a meaningful prediction. The logistic map allows you to do that only to the extent of the precision of your initial data. If that precision is limited, the predictability will also be limited. To add one additional meaningful time step of predictability requires an additional place of accuracy. That can be extraordinarily difficult to attain, depending upon the inherent precision of your measuring devices and computational tools.

How does all this relate to randomness? First, when the limits of the initial data are reached, the output of a computer is dominated by a combination of noise and a host of computational errors and approximations. The noise may in fact originate in quantum-mechanical processes within the electronic components and be legitimately random. Round-offs and computational approximations are probably not thoroughly random but can appear to be so. From a philosophical standpoint, the most important thing is that chaos places greater demands on precision than can be humanly achieved. The algorithm generated by the logistic map for $r = 4$ will produce a string of digits that are legitimately random by the definition of algorithmic complexity. This is not because of the complexity inherent in the process of multiplication by a power of two, but rather because of the complexity involved in generating the arbitrary string of digits needed as an initial value.

Why Does Chaos Look So Chaotic?

Many scientists have studied the chaotic orbits produced by the logistic map of $r = 4$. There has also been intensive study of the behavior of differential equations in chaotic regimes. One of the fascinating discoveries in the case of the map has been the fact that at $r = 4$—and presumably at other values of r as well—an enormous number of initial values between 0 and 1 do not lead to chaotic orbits but produce genuine periodic orbits. As enormous as this number is, however, so many more initial values produce chaotic orbits that a random choice of initial value will almost certainly produce a chaotic orbit. If you attempt to plot values that lead to periodicity and those that produce chaos, their intermingled ranges have a fractal structure. This is somewhat analogous to the fact that as r is varied but the initial value is held fixed, windows of periodic orbits appear within the sea of chaotic ones, containing the entire diagram. Thus periodic orbits appear to be interspersed with chaotic orbits in two different ways: for the same initial condition and different r values, and for a single r value and different initial conditions.

If you consider how easy it is to lose accuracy during a calculation of thousands of iterations, you can easily imagine that the periodic orbit you happen to have chosen might vary infinitesimally and throw you into a nearby chaotic orbit. In other words, you can find yourself meandering aimlessly, even though you may have selected an initial value that should have produced an endless repetition.

What Is Chaos After All?

We have now looked at chaos in both differential equations and discrete maps. We have seen that it has two characteristics. First, it is a form of behavior that appears to be random and without discernible order. Mathematically, one can say that this is due to the nature of the strange attractor, which is the limiting form of chaotic motion. Physically, it is a result of forcing, damping, nonlinearities, and the number of degrees of freedom, which combine to produce sudden and exaggerated changes in behavior with no apparent order or regularity.

Second, chaos is a form of behavior so extraordinarily sensitive to its initial conditions that its deterministic equations have no predictive value. What good is it to know you are on a unique orbit if you haven't got the necessary precision to know exactly which orbit you are on? You can't even draw an orbit with a pencil, because the thickness of the line at any point in phase space embraces a sufficient range of coordinate values to generate countless diverging orbits. If you sharpen the pencil point, you simply forestall the inevitable. No mesoscopic pencil is sufficiently precise to allow the drawing of a single orbit that extends for any useful distance. In short, chaos turns determinism into an abstract philosophical attribute. The useful attribute of predictability is lost. If you accept the definition of randomness as algorithmic complexity, then chaos produces genuine randomness. The chaotic production of points in a discrete logistic map is precisely what would be produced by an algorithm whose complexity exceeded that of the succession of points. The properties that make chaotic systems unpredictable also make the succession of states they produce totally random.

Twelve

Conclusion

I have now traced a path through some of the most important creations and discoveries of modern physics. As promised, I told how each has left us with new limits on the nature of knowledge, new ways of thinking about the physical world around us, and new forms of mathematical reasoning for describing the world. Special and general relativity are symmetries, defining things that we simply cannot know. In the case of special relativity, we cannot measure uniform movement through the universe; in the case of general relativity, we cannot distinguish between gravity and motion that is not uniform. Innocuous as these results may seem, their implications for a deeper understanding of nature are immense. The ignorance implied by symmetry is of such a high degree that it allows the "knowable" world to be described in the mathematics of geometry. In ancient times, mathematics was meant to describe the beauty of an ideal universe created by the gods. Something in the nature of geometrical forms and proofs echoed the presumed perfection of the transcendent world. Einstein resurrected geometry, adding curving spacetime, to describe a universe whose beauty is reflected in fundamental symmetries.

Statistical physics relates everyday phenomena to the unseen behavior of uncountable numbers of atoms and molecules that form the structure of matter. Once we realize that our apparently calm mesoscopic world hides a seething mass of molecular motion, we must also concede that the description of that world in terms of macrostates can never display the detail of a description in terms of microstates. We come to grips with this by giving up the possibility of ever describing mesoscopic behavior with certainty and accepting instead a probabilistic interpretation of our reality. This does no harm, except to our pride. The Newtonian dream of a completely predictable physical world must be replaced by a more modest dream of a world whose predictable behavior is at most probable. The importance of the concept of entropy to all of the sciences is evidence of our acknowledgment of the impossibility of accurately describing the behavior of enormous numbers of invisible entities. Entropy, after all, is a measure of order, complexity, and probability, a constant reminder of how much we cannot say about what we think we know is actually going on.

Quantum physics, like statistical physics, attempts to connect the microscopic and mesoscopic worlds, in this case the microscopic behavior of individual systems and mesoscopic measurements. Unlike statistical physics, the paradoxes that arise are the result not of our inability to describe enormous numbers of microscopic objects but rather of a fundamental discontinuity between the microscopic and mesoscopic worlds that exists even at the level of individual systems. We use measurements to build up a mental image of what the microscopic world is like, concluding that the systems in that world seem to be composites of mutually contradictory attributes. The way out of this situation is to resort to a completely probabilistic description of microscopic reality in which we deal, not with the system itself, but only with the probabilities of the results we will get when we observe it. To do this, we invent a new tool, the wave function, which contains all the information about the system that can be obtained by experimental observation. It also reflects the discouraging fact that the microscopic world is enveloped in a cloud of uncertainty. Pairs of variables can never be simultaneously measured with accuracy; experimental results appear randomly. The certainties of the mesoscopic world that we have become accustomed to seem to be the exception rather than the rule. The uncountable number of atoms of which we are composed obey the laws of quantum mechanics, not those of Newtonian mechanics. Why it is that we ourselves should be Newtonian is still a major mystery.

Finally we come to chaos, the newest challenge to the little bit of accurate knowledge and certainty remaining in our lives. Chaos proves that we must differentiate between the seemingly identical concepts of deterministic and predictable processes. It also provides us with a way of understanding the randomness that seems prevalent in nature yet has nothing to do with microscopic processes. Relativity and quantum physics may seem esoteric to the average person and appear to be relevant only to the practicing physicist in a laboratory, chaos has immediate and broad implications. The financial markets may be chaotic systems. The weather, about which we all talk but do nothing, is a chaotic system. Epidemics, the erratic beating of a diseased heart, and the cycles of defoliation by gypsy moths may all bear evidence of an underlying chaos. This newest field of study seems already to be playing an important and unifying role in our understanding of many phenomena that had been deemed totally unrelated. Yet chaos, like the other topics I have discussed, represents a significant diminution of what we can hope to predict about the course of natural events. While increasing our knowledge of phenomena, it has virtually destroyed the Newtonian and Laplacian dream of prediction and control.

Index

About the Author

MORTON TAVEL received his Ph.D. in physics from Yeshiva University in 1964. After three years at Brookhaven National Laboratory in the reactor theory and applied mathematics departments, he accepted a position at Vassar College in 1967, where he has been ever since. He was one of the founders of Vassar College's multidisciplinary program in science, technology, and society, which he directed for about ten years.